0293503·0

RESEARCH MONOGRAPHS ON HUMAN POPULATION BIOLOGY

General editor: G. AINSWORTH HARRISON

THE
GENETICS OF THE
JEWS

A. E. MOURANT

Formerly Director, Medical Research Council Serological Population Genetics Laboratory, London
Formerly Honorary Senior Lecturer in Haematology, St. Bartholomew's Hospital, London
Formerly Director, M.R.C. Blood Group Reference Laboratory, London
Sometime Visiting Professor of Serology, Columbia University in the City of New York

ADA C. KOPEĆ

Formerly Statistician, M.R.C. Serological Population Genetics Laboratory, London
Formerly Statistician, Nuffield Blood Group Centre, London

KAZIMIERA DOMANIEWSKA-SOBCZAK

Formerly Librarian, M.R.C. Serological Population Genetics Laboratory, London
Formerly Librarian, Nuffield Blood Group Centre, London

CLARENDON PRESS · OXFORD

Oxford University Press, Walton Street, Oxford OX2 6DP

OXFORD LONDON GLASGOW
NEW YORK TORONTO MELBOURNE WELLINGTON
IBADAN NAIROBI DAR ES SALAAM LUSAKA CAPE TOWN
KUALA LUMPUR SINGAPORE JAKARTA HONG KONG TOKYO
DELHI BOMBAY CALCUTTA MADRAS KARACHI

© *A. E. Mourant, A. C. Kopeć, and K. Domaniewska-Sobczak 1978*

ISBN 0 19 857522 X

British Library Cataloguing in Publication Data

Mourant, Arthur Ernest
    The genetics of the Jews.—(Research monographs
on human population biology).
    1. Jews   2. Human population genetics
    I. Title   II. Kopeć, Ada Christina
    III. Domaniewska-Sobczak, Kazimiera   IV. Series
    572.8'924   GN547   78-40437

ISBN 0-19-857522-X

*Printed in Great Britain
by William Clowes & Sons Ltd.
London, Beccles, and Colchester*

PREFACE

THIS book is an expansion of a paper published in 1959 on the blood groups of the Jews, which, initially somewhat reluctantly, I wrote at the request of my friend the late Professor Maurice Freedman. When, however, I found the initial paper growing considerably bigger than I had originally contemplated, I realized that it might later develop into something much bigger still. I therefore asked his permission, which he gave me, to incorporate parts of it into any future publication—one of the diagrams and some passages from the text have indeed been included in the second edition of *The distribution of the human blood groups* (Mourant *et al.* 1976a) and, with the permission of the present editor of *The Jewish Journal of Sociology*, rather more extensive extracts have been incorporated in the present book.

My own interest in the history of the Jews and of the Dispersion, however, goes back to early childhood for my first school teacher was, as I now realize, a British Israelite who believed that all the British people, if not all Europeans, were descended from the 'lost tribes of Israel'. I therefore believed that I, and all the people among whom I lived, were really Jews, and that the biblical prophecies about the return from captivity, and the eschatological prophecies about the Jews in The book of Revelation applied to us. Towards the end of the First World War, I thought these prophecies were at that very time being fulfilled in the liberation of Palestine by British armies, and the return of the Jews, shortly to be followed by the End of the World. Thus, for some of my most formative years, I considered myself a Jew, and this sense of identity has to some extent persisted and, though I can now see no evidence that I have any Jewish ancestors, I have maintained a deep interest in the Jewish peoples.

When *The ABO blood groups* (Mourant *et al.* 1958) was written, the data on the Jews were tabulated separately; this greatly facilitated the writing of my 1959 paper, and, partly with the present book in mind, the same scheme was used in the second edition of *The distribution of the human blood groups* (Mourant *et al.* 1976a).

This new book, is however, much more than a rearrangement of extracts from those tables. Many papers, both from my own research group and from others, on the genetic characters of Jewish populations, have since been published. Much more important, however, is the fact that when our Jewish friends knew that my colleagues and I were contemplating this book they unanimously, and with great generosity, sent us copies of very large quantities of unpublished data, so that, for several genetic systems, this new material exceeds in amount all that had previously been published. I hope that, before this book appears in print, much of this material will have been published independently by the original authors, in which case the appropriate references will, as far as possible, be incorporated in the bibliography of this book. Any work, however, which, as far as we are able to ascertain, remains unpublished, will be so acknowledged.

Most of the available data on the genetics of the Jews refer to blood groups in the restricted sense of red-cell agglutino-gens, and especially to those of the ABO and Rhesus systems, but there is a large and growing body of data on the other hereditary characters of the blood—the plasma proteins, the red-cell enzymes, the haemoglobins, and the histocompatibility antigens of the lymphocytes. Two other genetic systems—those controlling the ability to secrete the ABH blood-group antigens in the saliva and the ability to taste phenylthiocarbamide—have, traditionally and conveniently, always been considered by geneticists alongside the hereditary blood characters, and are so included here.

The genetic characters already mentioned include nearly all those which are of most use in the study of population genetics. There are, however, two other classes of inherited characters to which consideration must be given. One comprises the external characters of the body, and the other, the hereditary diseases.

In popular anthropology since time immemorial, and in a more precise manner for more than a century, the visible and physically measurable body characters preceded the hereditary blood characters as the principal human taxonomic markers. Because of the more precisely known mode of inheritance of the blood characters, the latter have now largely superseded them, but the visibly expressed characters, and especially those of the skeleton, provide the only means for comparing living populations with those of the historic and prehistoric past, and with those samples of populations still living, which were examined and described by physical anthropologists before the discovery of the blood groups. Morphological studies lie outside the competence of the present authors but an attempt will be made to summarize the relatively little which is known of the comparative anthropometry of Jewish populations.

The hereditary diseases are, in general, too rare to be of value in studies of limited population samples, but there is a considerable literature, much of it anecdotal rather than precise, on the relative incidence of particular diseases in Jewish and non-Jewish populations. Some of these diseases, such as Tay–Sachs disease, are inherited in a precisely known manner, but others are probably (like the visible characters of the body) the results of the interaction of a number of independent genes, and their incidence can be described only as familial. However, much has been done recently to define the genetics of such diseases, especially by showing them to have statistical relations with well-defined blood characters and, especially, in very recent years, with the histocompatibility antigens. The present authors (Mourant *et al.* 1978) have recently made a comprehensive study of such associations.

Much of the literature on diseases in Jews refers to Askenazi Jews, compared with other west and central Europeans, and a large part of it is the work of Jewish doctors. There is now, however, in Israel, an opportunity for comparisons with Jews immigrating from countries outside Europe. A most useful collection of reprints of leading papers on diseases in Jews has been made by Shiloh and Selavan (1973) and we have consulted it extensively. An excellent but now largely outdated book on the Jews of the world is

The Jews: a study of race and environment by M. Fishberg (1911). It contains a very full account, in contemporary terms, of their physical anthropology, including numerous photographs.

In the preparation of this book, the bibliographic work has been primarily the responsibility of K. Domaniewska-Sobczak, who also, especially in the case of data transferred from *The distribution of the human blood groups,* carried out a large part of the searching of the literature. The statistical part of the work has been the responsibility of A. C. Kopeć, who extracted nearly all the data, either expressly for this book, or for our previous one, from the original publications and from unpublished typescripts. She also carried out the computations or, in some cases, prepared the data for electronic computation. She then prepared the tables for publication from the computation results.

The introduction to the tables is by A. C. Kopeć and A. E. Mourant. This preface and the remainder of the text have been written by A. E. Mourant with much consultation with his co-authors.

I am greatly indebted to my wife, who typed the text, some of it many times over, from my untidy manuscript.

The map was prepared from a list of place-names, and the diagrams drawn from my drafts, by Mr. J. Hunt, with the support of a grant from the U.S. Army Research and Development Group (Europe). The work as a whole was carried out with the support of a grant from the Wolfson Foundation, in accommodation kindly provided by St. Bartholomew's Hospital, London. We were also accommodated, in the last stages of the work, in the Anthropology Sub-department of the British Museum (Natural History), London.

We are much indebted to the librarians and staffs of numerous libraries, especially those of the Royal Society of Medicine, the Royal Anthropological Institute, the Medical College of St. Bartholomew's Hospital, and the Department of Hebrew Studies, University College, London. I am grateful for helpful criticism and encouragement received from Professor C. Abramsky and the Rev. Dr. J. R. Rosenbloom.

We are grateful to the University of Toronto Press for permission to publish the poem on page 26.

London, 1977   A.E.M.

CONTENTS

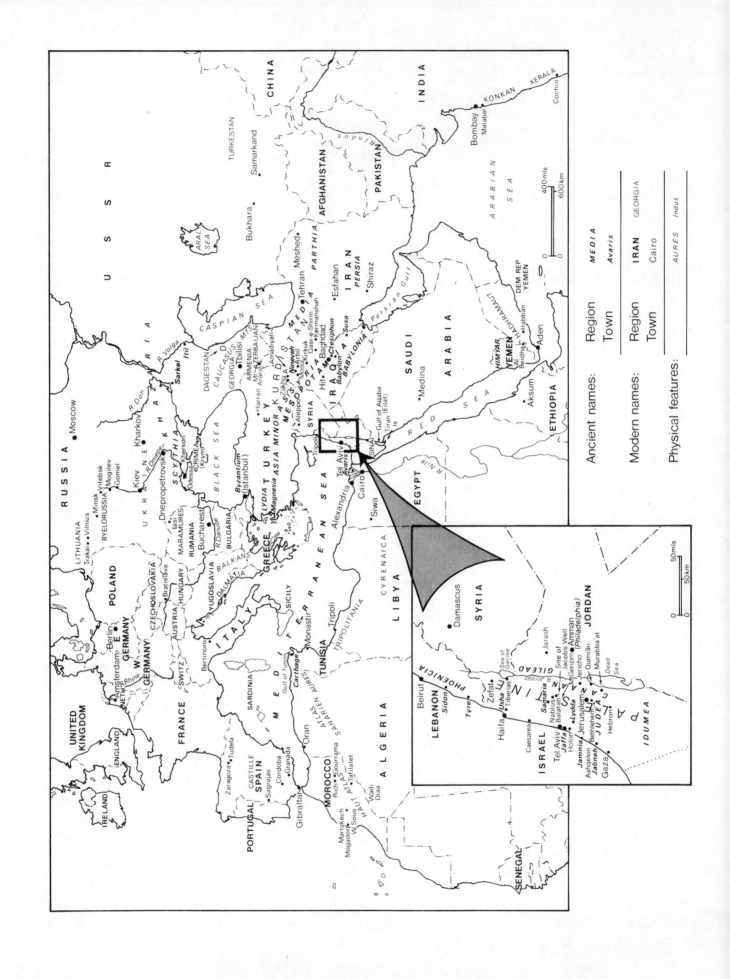

1

INTRODUCTION

IT is generally recognized that the Jews of the Dispersion show both cultural and physical differences from the various populations among whom they live, or among whom they lived until they went to Israel. Opinions vary, however, as to the relative importance of these two kinds of difference. There is, moreover, a divergence of opinion not only as to the extent to which even the physical differences are environmental rather than inherited, but also as to how far inherited differences are derived from an original Israelite ancestry, how far from intermarriage in lands of temporary sojourn, and how far from the incorporation into the Jewish community of non-Jewish populations who have been converted to Judaism. It has also been suggested that there has been genetical self-selection of persons leaving the Jewish community through marriage or for other reasons, but it would be surprising to find that any such process had had any measurable effect on the frequencies of the blood groups or other genetic characters in the remaining Jewish population.

Before we can make specific comparisons between the Jews and their neighbours we must consider in general terms the nature of the evidence available for comparing populations with one another. Whereas my earlier paper (Mourant 1959) was particularly concerned with the comparison between Jews and non-Jews, here we shall consider also the genetical characteristics of Jewish populations in their own right, comparing them with one another, as well as with their Gentile neighbours.

As already mentioned, physical anthropologists, until a few decades ago, depended for the comparison of populations almost entirely on observations and measurements of the anthropometric and mainly external characters of the body. Differences in such characters are obvious to all observers, but unless observations of them are made in a strictly quantitative and controlled manner, they are liable to be coloured by personal prejudices: we each have our own mental picture of what a Frenchman or a Japanese should look like, and are liable to see this picture in every Frenchman or every Japanese.

Objective differences between individuals and between populations can, however, readily be demonstrated if observations of the size, shape, and colour of the body and its parts are made which are strictly quantitative, but even these will not tell us whether the differences are due to environment or to heredity. The basic characteristics are certainly inherited, but the manifestation of the genes responsible may be considerably modified by the environmental history of the individual. Also, the genetics of the visible characters have been very little studied, though it is certain that each of them, with one or two possible exceptions, is the result of the joint operation of a considerable number of genes at different loci.

The blood groups and other inherited blood factors, on the other hand, are genetically relatively simple and well understood. The observed facts, that is to say the results of our tests on the blood of the individual, constitute his or her blood phenotype. For each genetic system the latter is very directly and simply related to the genotype, or genetical constitution, which is fixed for life at the moment of conception. As far as most tests are concerned the phenotype too is fixed by the time of birth, and for all tests by the age of about one year. Unlike the visible characters, the inherited blood characters are therefore unaffected by the environment. They have the further practical advantage that their consideration is unaffected by such emotional accretions as have tended to attach themselves to discussion of the visible features of the body.

When my 1959 paper on Jewish blood groups was written, the number of known systems of hereditary blood types was still small; only a few of these systems had been applied to human classification, and still fewer to the characterization of the Jews, or of individual Jewish communities. These few were, almost exclusively, systems of blood groups in the restricted sense of surface antigens on the red blood cells.

The number of known systems of hereditary blood factors is now very large, probably running into many hundreds, of which nearly one hundred have been involved in population studies, and some thirty in such studies of Jewish populations. These comprise, besides the classical blood groups, numerous systems of plasma proteins and of red-cell enzymes, as well as the haemoglobin system, and the very complex system of lymphocyte and tissue antigens, the histocompatibility or HLA system. Two other polymorphic systems are particularly involved in population studies, those concerned respectively with the secretion into the saliva of the antigens of the ABO blood-group system, and with the ability to taste phenylthiocarbamide.

For all the polymorphisms covered in the present book, full descriptions of world distribution and of population genetics will be found in a previous book by the present authors (Mourant *et al.* 1976*a*). A comprehensive account of the genetics and other aspects of the blood groups in the strict sense will be found in the work of Race and Sanger (1975). The blood protein factors are covered by Harris (1970) in *The principles of human biochemical genetics. Genetic markers in human blood* by Giblett (1969) deals comprehensively with all the blood factors with which we are concerned, and in particular gives a full account of the immunoglobulins and some other serum factors which fall outside the scope of both the other books. Lehmann and Huntsman (1974) have written a comprehensive account of the haemoglobins while Livingstone (1967) has described and tabulated their world distribution and Professor A. G. Steinberg is preparing a similar monograph on the immunoglobulin variants.

The books mentioned deal, up to varying levels of sophistication, with general genetics and the genetics of the particular systems described, but many readers may feel the need of a deeper treatment of this subject such as may be found in the *Principles of human genetics* by Stern (1973). *The assessment of population affinities in man* (Weiner and Huizinga 1972) deals with the subject of its title much more fully than can be attempted here.

In the writing of this book, as in that of our more comprehensive one, we have been faced with the problem of how to set out observational findings. For most purposes the results of the examination of the blood factors of a population sample are expressed in the first instance as a series of numbers each representing the total of individuals falling into a given serological class. Each of such classes, not further divisible by means of the reagents and facilities available, constitutes a phenotype. Within each of such classes or phenotypes, however, and especially in the case of the blood groups in the restricted sense, there may be individuals who are of several different genetical constitutions or genotypes; rather rarely in the blood group systems, but commonly in the blood protein systems, each genotype constitutes a separate phenotype. Where a phenotype is composite it remains a provisional entity which may later, or in another laboratory, be divisible into a number of new phenotypes which may, however, still be composite. Even in those systems where it is not possible to determine the genotypes of all individuals by direct tests, we can, however, nearly always calculate with a high degree of probability the frequencies of the genes, and from them the frequencies of the genotypes, in the population as a whole.

It might at first sight appear simpler in considering, for example, the ABO blood groups to express and discuss all results in terms of the observed phenotype frequencies. We should then have the satisfaction of using numbers representing something observable in individuals, at the expense of a slight increase in complexity—having to deal with four variables instead of three. Indeed, in a few systems, though there are three genotypes, there are only two phenotypes and two genes, the probable frequencies of the genes bearing a simple one-to-one relationship to the observed frequencies of the phenotypes. In the more complex systems such as Rh, however, the number of phenotypes becomes unmanageable and one can only think clearly of results in terms of the frequencies of the genes (or in the particular case of Rh, of the gene complexes).

There is, however, a more compelling reason for using gene frequencies. Not only are the genes the more fundamental units, but in considering mixing of populations, and such phenomena as mutation, selection, and frequency drift in small populations, it is necessary to think and to calculate in terms of genes. Mixing alone is such an ever-present consideration that gene-thinking must become second nature to anyone who attempts to apply genetics to anthropology or to population problems of any kind. For these reasons the discussions in this book are in general expressed in terms of genes rather than of phenotypes.

The greater part of this book consists of a succession of studies, each dealing with one main Jewish population or group of populations. Each begins with a summary of what is known of the history of that population; this is followed by a consideration of the frequencies which it shows for each of the various blood factors, and an attempt is then made to relate these to one another.

Where information is available on the anthropometry, or any particular physiological characteristics, these will be mentioned. The object here will be to direct readers to original sources of information rather than to provide a critical account. Most of the published data are old and do not conform to modern standards. There is a great need for a comprehensive and up-to-date anthropometric survey of the Jews in the world.

A separate chapter deals with the hereditary predispositions of Jews in general, or of certain Jewish communities, to suffer from, or to be free from, particular diseases.

2

BLOOD GROUPS AND OTHER POLYMORPHISMS

IN the previous chapter, references have been given to comprehensive accounts of human genetics, and of the blood groups and other polymorphisms which have been used in population studies. This chapter provides only a condensed account, for immediate reference, of those systems of blood groups and other blood factors for which Jewish populations have been tested. In writing these descriptions a previous knowledge of at least the elements of genetics has been assumed. It has not been thought necessary to give a bibliographic reference for every statement made in this chapter, but such references can be found by consulting our previous book (Mourant *et al.* 1976a). The present account, though placed early in the book, is intended rather for reference than for continuous reading.

THE ABO BLOOD GROUPS

The phenotypes known as the ABO blood groups are determined by a system of three allelomorphic genes *A*, *B*, and *O*. Of these, *A* and *B* each gives rise to a characteristic antigenic (and ultimately biochemical) structure, the A or the B antigen, while the *O* gene behaves as an amorph, not giving rise to any antigen peculiar to itself. The usual tests for these antigens detect them on the surface of the red cells, but they are in fact found widely distributed elsewhere in the body.

Because the *O* gene behaves as an amorph, not giving rise to any specific gene product so far recognized, the genotypes *AA* and *AO* are indistinguishable by any routine tests on the cells of the individual, as are the genotypes *BB* and *BO*.

The standard reagents for the A and B antigens are human sera containing the specific antibodies anti-A and anti-B. Each of these causes the agglutination of red cells carrying the corresponding antigen. All normal human sera contain such of these antibodies as do not react with the individual's own red cells. Thus the serum of an A person contains anti-B, that of a B person, anti-A; the serum of an O person contains both antibodies, and that of an AB person, neither of them. The relations between genotypes, phenotypes, and antibodies are summarized in Table I.

The A and B antigens result from the action of enzymes, glycotransferases, which catalyse the conversion of a common basic antigen, known as H, into A or B. The H antigen itself is, indirectly, the product of a gene *H*, whose amorph

allele is *h*. *H* and *h* are not linked to the *A* and *B* genes. Extremely rare individuals, particularly in and near Bombay, lack the *H* gene and are of genotype *hh*. Thus, whether or not they possess an *A* or *B* gene, their red cells lack the A or B antigen. They are recognized by the presence in their plasma of the antibody anti-H. No Jews have so far been recognized as belonging to the Bombay or *hh* type, but it is necessary to mention it here as it enters into theoretical discussions later in this chapter.

The sub-groups of A

Considerable numbers of variants of the A antigen are known, most but not all of which are rare; the B antigen is less variable but several variants, all very rare, are known. The most important sub-group distinction is that between A_1, the commonest antigen, and A_2, which has a frequency of several per cent in most European, African, and West Asiatic populations. Although the distinction has been known for nearly fifty years, its basic nature is still not completely understood, but most of the facts are covered by the following conventional account. Thomsen, Friedenreich, and Worsaae (1930) observed that there were two varieties of the A antigen, A_1 and A_2, allowing the blood groups A and AB to be classified respectively as A_1 and A_2, and as A_1B and A_2B. Both types of antigen react with the ordinary antibody anti-A, but only A_1 reacts with anti-A_1, while A_2 fails to do so. Reaction of antigens with antibodies is shown, as usual, by agglutination. Anti-A_1 is present in the serum of most B persons together with ordinary anti-A. The latter antibody can be absorbed out from a serum containing it, by means of A_2 cells, leaving only anti-A_1 behind, so that the serum becomes a specific anti-A_1 reagent. An excellent anti-A_1 reagent can also be prepared by extracting *Dolichos biflorus* seeds with physiological saline.

The A_1 and A_2 antigens are produced by corresponding allelomorphic genes, so that what we have called the *A* gene is really of two possible kinds, A_1 and A_2. In the genotype A_1A_2 the A_1 gene causes the production of A_1 antigen, and thus the genotypes A_1A_2, A_1O, and A_1A_1 are indistinguishable, since all react both with anti-A and anti-A_1. Similarly the genotypes A_2A_2 and A_2O are indistinguishable, both reacting with anti-A but not with anti-A_1. A few population samples were tested also for the intermediate type, A_{int} (see Mourant *et al.* 1976, pp. 4–6). When population samples have been tested for the subgroups of A, the results, classified only for the main four ABO groups, are included in Table 1, but are marked to show that they have been subgrouped and that the full results are given in another table.

A and *B* gene frequency diagrams

Various statistical and graphical methods have been devised to show the genetical distances between populations, derived from the frequencies of a range of marker genes. These however depend upon the availability of sets of frequencies of

TABLE. I. THE ABO BLOOD GROUPS

BLOOD GROUP (PHENOTYPE)	BLOOD-GROUP SUBSTANCES ON RED CELLS	ANTIBODIES PRESENT IN PLASMA (OR SERUM)	GENOTYPE
O	none	anti-A, anti-B	*OO*
A	A	anti-B	*AO* or *AA*
B	B	anti-A	*BO* or *BB*
AB	A and B	none	*AB*

genes, the same for all the populations to be compared. In the case of the Jews, the only genes of which frequencies are available for a sufficiently large number of communities are the *A*, *B*, and *O* genes of the ABO blood-group system.

The frequencies of these three genes must add up to 100 per cent, so that only two independent variables exist, and thus a complete representation of the ABO constitution of a population can be given by a point on a two-dimensional graph.

Various methods have been devised for this purpose, of which the most logical one is the Streng diagram with three axes at 120° from one another, along which are plotted the frequencies of the *A*, *B*, and *O* genes. The inclination of the axes to one another, however, makes these diagrams difficult to read by persons accustomed only to rectangular Cartesian graphs.

Thus, though they are theoretically less satisfactory, the easiest diagrams to read are those which represent the frequencies of the *A* and *B* genes along two axes at right angles. Each population is then represented by a point at the appropriate distances from the two base lines. The frequencies of the *O* gene can also be read if a third axis is inserted bisecting the angle between the other two.

In the diagrams in this book (Figs. 2–7) only the *A* and *B* axes are shown. In all the diagrams the *A* and *B* gene frequencies of any given Jewish population are shown by a single solid black spot while, in all except Fig. 7, the gene frequencies of the corresponding indigenous population are shown by an open circle. For each population the spot and the circle are joined by a line, the length of which is a measure, though not a precise one in the mathematical sense, of the genetic distance between the two populations.

These diagrams need to be examined in terms of a number of alternative hypotheses. If the Jews had remained a single homogeneous group, not intermarrying with indigenous peoples, then there would be one spot, in the same position in all diagrams, representing the Jews, and surrounded by a more or less random constellation of small circles representing the indigenous peoples.

If, at the other extreme, intermarriage and proselytization had rendered the Jews identical genetically with the autochthones, there would be only one point for each country, representing both Jews and natives.

If in each country there had been some degree of mixing and of proselytization, then, if there had been no complications such as natural selection or genetic drift, each Jewish population would be represented by a point on the line joining the small circle representing the indigenous population to a single point, the same in all cases, representing the composition of the 'original Jews'.

The actual diagrams are all much more complicated than any of these ideal ones, but several of them show a certain resemblance to the last described type. In the particular case of northern and central Europe (Fig. 5) there is a clustering of the Jewish populations around the frequencies *A*, 27 per cent; *B*, 12 per cent, with a slight tendency for the lines to radiate from this centre to the circles representing indigenous peoples.

THE ABH SECRETOR SYSTEM

Some persons do and others do not secrete into their saliva (and certain other body fluids) antigens corresponding to their ABO blood group, and the ability to secrete behaves as a simple Mendelian factor, dominant to non-secretion. Group A, B, and AB persons who are secretors secrete the antigens corresponding to their blood groups. Group O persons who are secretors secrete the H substance, which is the chemical precursor of the A and B substances, as do all other secretors to a somewhat lesser extent.

THE LEWIS SYSTEM OF ANTIGENS

Because of its close serological and biochemical association with the ABO blood groups and with secretion, the Lewis system (Mourant 1946) must be mentioned here. The Lewis antigens are essentially water-soluble antigens present in the saliva and other secretions, and in small quantities in the blood plasma, whence they are taken up by the red cells. They are the products of a pair of allelic genes hitherto known as *Le* and *le*. This notation wrongly implies that, as was believed for many years, *le* is an amorph. This is now known not to be the case, but the notation will be retained here provisionally, so as to avoid burdening the literature with what might prove to be an unacceptable new symbol. The primary product of the *Le* gene is the Le^a antigen, but in the presence of the secretor *Se* gene it is converted into the Le^b antigen (the presence of an *H* gene—see above—is necessary for the synthesis of the Le^b but not of the Le^a antigen, but as absence of *H* is exceedingly rare this is mainly of theoretical interest). The primary product of the *le* gene is the Le^c antigen (Gunson and Latham 1972) while the joint product with the *Se* gene is Le^d (Potapov 1970). Because of the many interconnections between the systems just mentioned, and the weakness of most of the available reagents, the establishment of the Lewis genotype of an individual is a matter of some technical difficulty, so that many of the published data on frequencies in populations must be accepted with some reserve.

The most reliable way to determine the Lewis type of an individual is by means of tests on the saliva, but most published results, though only one set for a Jewish population, concern the Le^a reactions of red cells, which are positive in persons who have at least one *Le* gene but are homozygous (*se se*) non-secretors of the ABH antigens.

THE MNSs BLOOD GROUP SYSTEM

The MN blood groups (Landsteiner and Levine 1927) depend upon a pair of allelomorphic genes which behave in nearly all respects as though they determined the antigens M and N respectively. It has however recently been shown that the N antigen is a precursor substance, and that the so-called *N* gene is an amorph which leaves the N antigen unchanged, while the *M* gene in the heterozygote converts part of the N antigen into M, and in the homozygote converts nearly but not quite the whole of it.

The S and s antigens are the products of a pair of allelic genes very closely linked to those for M and N. A third allele, which determines the absence of both S and s, and is known as S^u, is not uncommon in Negroids.

The system is one of considerable complexity involving numerous variants of M and N, and a variety of antigens determined by other closely-linked genes. The only two such antigens for which we can find records of tests on Jewish populations are the M^g variant of M, and the He (Henshaw) and M^t (Martin) antigens, which are the products of closely-linked genes. The M^t antigen is very rare in all populations, and of little anthropological significance, but the He antigen is a very useful marker of African ancestry.

A large number of Jewish populations have been tested

only for the M and N antigens, in addition to those tested for these and for S with or without s. To facilitate overall comparisons, Table 2.1 includes the results of all tests for M and N, whether or not additional tests were done, but those sets of data (marked with the appropriate symbol in the MN table), which include results of such additional tests as well, are again tabulated so as to show the complete test results [Tables 2.2 and 2.3].

THE P BLOOD-GROUP SYSTEM

The P blood groups were discovered by Landsteiner and Levine (1927) in the course of the same investigations that defined the MN groups. It was at first thought that only one antigen P was involved, determined by a gene P, the allele p being an amorph, the two genes having each a frequency of about 50 per cent in European populations. Further investigation has disclosed a system of considerable complexity, only one feature of which will be described here. The antigen Tj^a found by Levine et al. (1951) was at first regarded as the product of a gene present in nearly all human beings, the extremely rare allele being an amorph, with the homozygotes usually having a strong anti-Tj^a antibody. Sanger (1955) then showed that Tj^a was part of the P system, and that this system really comprises three alleles, P_1 (formerly P), P_2 (formerly p), and the new p (formerly regarded as the amorph allele of Tj^a). The relationships are similar to those existing between A_1, A_2, and O of the ABO system. P_2 bloods sometimes show anti-P_1 in the plasma, usually with a very low titre, but the rare p bloods always have a high titre of anti-P + anti-P_1. Although many tests have been done on populations, the results are not of great value for comparative purposes, as some series of data are believed to include false negatives.

THE RHESUS BLOOD GROUPS

From a clinical point of view the Rhesus or Rh system is by far the most important of the blood-group systems other than ABO. Rh incompatibility is the main cause of haemolytic disease of the newborn, and a major cause of transfusion reactions. The principal antigen of the system, Rh$_0$ or D, was discovered by Landsteiner and Wiener (1940). Subsequently, numerous other associated antigens were discovered, and considerable controversy arose both as to notation and as to the theoretical interpretation of the uncontroversial facts observed by serologists. This is not the place to discuss the controversy and for the purpose of this book the CDE notation of Fisher (Race 1944) will be used, and its genetical implications of closely-linked genes will be assumed. Only the main features of this very complex system will be described. On this basis the principal antigen of the system is known as D, determined by a gene D, the allele d of which behaves as an amorph. Very closely linked to the Dd locus are two other loci each characterized by a pair of major alleles, Cc and Ee respectively. Each of these four genes gives rise to a correspondingly named antigen.

There are thus eight possible chromosomic combinations of genes, all of which are known to exist, but of which CDe, cDE, cDe, and cde are the most common. The combination cde is mainly confined to Caucasoid populations but also occurs in Negroids, while cDe, present at low frequencies in most populations, reaches frequencies of 50 to 90 per cent in most African ones.

A number of rare alleles of C and c, of D and d, and of E and e exist. Among these is C^w, usually in the combination C^wDe, which is found with a frequency of about 1 per cent in most European populations but reaches higher frequencies in the north, especially the Lapps. Weak forms of the D antigen, collectively known as D^u, are present in low frequencies in most populations. The gene complexes CD^ue and cD^uE are present in low frequencies in most European populations, while cD^ue is typically African, with frequencies in Negroid populations up to about 10 per cent.

The V antigen is an important marker of African ancestry but, unlike most such markers, its highest frequencies are found in north-east Africa. Its genetical status is somewhat controversial but the gene or genes which give rise to it are undoubtedly part of the Rh complex, and are almost invariably linked to one of the combinations, cde, cD^ue, cDe. The Gonzales (Go^a) antigen is a part of the Rh complex and appears to be a variant of D. It seems to be restricted to Negroid populations, in whom it has frequencies of 1 to 4 per cent.

The results of Rhesus tests are shown in a series of tables, according to the particular combinations of tests performed. The derived gene frequency results shown are in most cases the frequencies of the threefold gene complexes (sometimes called alleles or haplotypes) such as CDe, cde. However, in all cases where tests for the D antigen were performed, the results of such tests are included also in Table 4.1, together with the results of tests for the D antigen only.

THE LUTHERAN BLOOD-GROUP SYSTEM

The Lutheran system depends upon a pair of allelmorphic genes Lu^a and Lu^b, which give rise to the corresponding antigens, which react with antisera containing the respective antibodies, anti-Lu^a and anti-Lu^b. With the use of these two antisera three phenotypes can be identified, each corresponding to one of the genotypes Lu^aLu^a, Lu^aLu^b, and Lu^bLu^b. A forth phenotype is known which fails to react with either serum. Most examples of this phenotype show a dominant inheritance possibly due to a gene at a separate locus, but no such example has ever been found in a random population survey. Others show a recessive inheritance and may be due to an amorph allele at the Lu locus; only three examples have been found in population surveys.

Both the diagnostic sera required for this system are in short supply, anti-Lu^b extremely so. Thus only a very few population samples have been tested with both antisera, and only moderate numbers with anti-Lu^a alone.

The Lu^a gene has a frequency in persons of northern European origin of about 4 per cent. In the Mediterranean area the frequency falls to about 2 per cent, but in Africa, though frequencies are variable, the average is nearly as high as in northern Europe. The gene is very rare or absent in all other indigenous populations examined.

THE KELL BLOOD-GROUP SYSTEM

The Kell blood-group system which initially, like all other systems, appeared simple, has gradually been shown to have a comparable complexity, and a similar organization, to the MNs and Rh systems. The Kell antigen was first described in 1946 by Coombs, Mourant, and Race. It was soon shown that it was the product of a gene K; and an antibody recognizing the product of the allelic gene k was discovered by Levine et al. (1949).

Two further antigens, Kp^a and Kp^b, were discovered by Allen and Lewis (1957) and Allen et al. (1958). They were shown to be the product of a pair of allelic genes associated with the Kell system.

The Sutter system of Giblett (1958) at first appeared to be an independent one. Again two antithetical antigens, Js^a (found by Giblett) and Js^b (found by Walker *et al.* 1963), are known, the products of a pair of allelic genes, first shown by Stroup *et al.* (1965) to belong to the Kell system.

Finally, the Karhula or Ul^a antigen of Furuhjelm *et al.* (1968) has been shown to have a similar association.

By analogy with the MNSs and Rh systems, the Kell system may be regarded as determined by three or possibly four closely-linked loci, occupied by the respective allelic genes K, k; Kp^a, Kp^b; and Js^a, Js^b. The details of the relationship with Ul^a are not yet published.

If we disregard Ul^a, about which little is yet known, only four of the eight theoretically expected gene complexes have been found, as shown in Table II. Some very rare individuals are known whose red cells are negative with all six diagnostic antisera. They may be homozygous for a rare amorph gene complex, but it is possible that some of them are, as suggested by Race and Sanger, homozygous for a rare allele of a common gene which controls the synthesis of a basic substance from which the observed antigens are elaborated.

TABLE II. GENE COMPLEXES OF THE KELL SYSTEM

KKp^bJs^b
kKp^aJs^b
kKp^bJs^a
kKp^bJs^b

The great majority of human beings in all parts of the world are homozygous kKp^bJs^b/kKp^bJs^b, the genes K, Kp^a, and Js^a being everywhere relatively rare and no two (or three) of them having yet been found occurring together in one gene complex. Thus most of the facts about the distribution of the complexes can be covered by separate consideration of the distributions of the genes K, Kp^a, and Js^a.

The K gene is found mainly in Caucasoids. Its frequency in European populations is usually between 3 and 5 per cent, but somewhat higher in parts of Scandinavia, though very low among Lapps. The highest known frequencies of the K gene are found among peoples of the Arabian and Sinai Peninsulas, in whom it often exceeds 10 per cent. It is, however, rare in the rest of south-west Asia and in Africa, and almost totally absent in the Mongoloid and other peoples of eastern Asia, the Pacific area, and America. It is present in varying frequencies in most Jewish populations.

The Kp^a gene has a frequency around 1 per cent in populations of European origin. Little is known of its distribution in other peoples.

The Js^a gene is almost entirely confined to the peoples of Africa and of known African ancestry. Next to the V antigen of the Rh series, the Js^a antigen is potentially the most valuable indicator of African admixture. The Bedouin of the Sinai Peninsula are among the few known populations having relatively high frequencies of both K and Js^a genes; if, therefore, these genes are, as here assumed, on closely adjacent but separate chromosome segments, it is in this area if anywhere that one might expect occasional crossing over to occur, giving rise to a gene complex containing both K and Js^a.

The Ul^a gene has so far (apart from one Swede and one Chinese) been found only in Finns in whom it has a frequency of 1·3 per cent, but up to 2·5 per cent in some isolates.

THE DUFFY BLOOD-GROUP SYSTEM

The antigen known as Fy^a was discovered by Cutbush *et al.* (1950) and the Fy^b antigen, the product of its allelic gene, by Ikin *et al.* (1951). The two genes Fy^a and Fy^b account for nearly all the phenotypes found in European populations, but it was observed by Sanger *et al.* (1955) that a high proportion of American Negroes are of the phenotype Fy(a−b−). An inspection of the latest tables of Mourant *et al.* (1976a) will show that, except in the South African Republic, over 95 per cent of African Negroes are of this type, which represents the homozygote of the third allelic gene, at first regarded as an amorph and so named Fy. It is now known that two distinct genes have in the past been confused under this term. One, Fy^x, with a frequency of approximately 1·6 per cent in Europeans, gives a product which is negative with anti-Fy^a but which reacts feebly with anti-Fy^b (Lewis *et al.* 1972). The other, almost universal in Africans, gives a product which completely fails to react with either of these antibodies, but is not a true amorph, for Behzad *et al.* (1973) have shown it to react specifically with an antibody which they call anti-Fy^4; the gene should presumably be called Fy^4. This gene is thus a valuable marker of African ancestry, but its detection and the determination of its frequency have hitherto depended upon the use of the two sera, anti-Fy^a and anti-Fy^b. Since the latter is usually a very weak reagent, false negative results may have led to exaggerated estimates of the frequency of the double-negative phenotype and hence of the corresponding gene. If supplies of the positive reagent anti-Fy^4 should become available, much more confidence could be placed on estimates of the gene frequency in certain populations of the Near East, including some Jewish ones.

THE KIDD BLOOD-GROUP SYSTEM

The Kidd blood-group system appears to be a very simple one, depending upon a pair of allelic genes Jk^a and Jk^b, giving rise to a corresponding pair of antigens, detectable by the use of a corresponding pair of antisera. The latter are rather scarce, so that relatively few populations have been tested either with one or both of them. This is unfortunate, as the gene frequencies span a useful range, that of Jk^a being about 75 per cent in Africans, 50 per cent in Europeans, and 30 per cent in Chinese. The phenotype Jk(a−b−) is known but is extremely rare and does not appear to be confined to any particular racial group.

THE DIEGO BLOOD-GROUP SYSTEM

The Diego antigens are the products of a pair of allelic genes Di^a and Di^b, producing a corresponding pair of antigens. The Di^a antigen is almost entirely confined to persons of Mongoloid origin. The Di^a gene has a frequency of about 4 per cent in most central and eastern Asiatic populations, but it reaches higher levels in Amerinds, especially those of South America. It is found, in very low frequency, in the Arabs of the Hadhramaut, having apparently been introduced as a result of the centuries-old traffic between that area and south-east Asia.

THE DOMBROCK BLOOD-GROUP SYSTEM

The Dombrock system is defined in terms of the one antigen, Do^a, which is at present detectable with the one type of antiserum at present available (and very rare). This antigen is the product of a gene *Do^a* to which corresponds an allele which, since it has so far behaved as an amorph, must provisionally be called *Do*, without an index letter. The frequency of the *Do^a* gene among northern Europeans is about 42 per cent, and it is near to this figure in white Americans and in a mixed sample of Jews now in Israel, mainly from Iraq. It is about 30 per cent in a small sample of American Negroes, so it may be lower still in Africa. Rather surprisingly, it appears to be 34 per cent in American Indians from the United States but only 7 per cent in Thais.

THE Xg BLOOD-GROUP SYSTEM

Genetically the Xg system is one of the most important of the blood-group systems, and population studies would be of great interest, but relatively few have been done, since the small available supplies of the one type of antiserum, anti-Xg^a, have been devoted mainly to fundamental genetical investigations, many of them carried out on Jewish families in Israel.

It was, however, a population study by Mann *et al.* (1962) which not only defined the newly discovered antigen Xg^a but showed that the gene determining it was on the X chromosome, for among Caucasoids from England, Northern Europe, and America there were 64·6 per cent positives among males but 89·3 per cent among females, frequencies which correspond closely to those expected for the product of an X-linked gene (*Xg^a*), with a frequency of 65·9 per cent. The gene frequency is the same as the phenotype frequency if males alone are considered, but male and female frequencies may be combined to give a more precise gene frequency estimate. The X-linked inheritance has been abundantly confirmed by family studies, since the Xg system has been used very extensively both in the mapping of the X chromosome and in investigating the cytological origin of sex-chromosome abnormalities and their anatomical and physiological manifestations.

Frequencies of the *Xg^a* gene in European populations show a range of frequencies from 76 per cent in Sardinians to 50 per cent in Finns and Norwegians. Frequencies in Chinese and Japanese are somewhat lower than in Europe, and the Tayal of Taiwan have only 38 per cent, but North American Indians, and the Aborigines of Australia and New Guinea all have frequencies in the neighbourhood of 80 per cent.

THE WRIGHT AND RADIN BLOOD-GROUP SYSTEMS

The Wright (Wr^a) and Radin (Rd) antigens are both extremely rare in all populations but because of the availability of relatively large amounts of the diagnostic antisera they have been applied extensively to population studies.

The Wright antigen was found by Holman (1953). Though it is very rare, the early date of its discovery and the ready availability of testing sera have led to very large numbers of persons being tested. One other reason for many of the tests is that a great many testing sera for common antigens are known to contain anti-Wr^a as well, and these sera have often been used in parallel with a known pure anti-Wr^a serum, to avoid false positive results arising from the use of the mixed serum. The frequency of the *Wr^a* gene in Europeans is about 3 in 10 000.

The Radin (Rd) antigen was discovered by Rausen *et al.* (1967) who found three persons possessing it among 562 New York Jews, but none among over 6000 other people. Lunds-

gaard and Jensen (1971) have, however, found 62 among 14 301 Danes, not known to be Jewish, and Bjarnason *et al.* (1968) one among 529 Icelanders. It thus appears to be slightly less rare among Jews and Scandinavians than among other peoples.

THE PLASMA PROTEINS

The blood plasma contains in solution a great variety of proteins, many of which show genetical polymorphism. Unlike the blood groups, these proteins mostly have known functions, for instance as enzymes, or as carriers of simple substances like metals and vitamins. The genetical variants of a particular protein often, and perhaps nearly always, differ quantitatively in their functional activity, thus allowing the possibility of natural selection, which may be a major cause of the observed variability of gene frequencies between populations.

As in the case of the blood groups, only those systems will be described which have been applied to the study of Jewish populations.

The haptoglobins

The haptoglobins are glycoproteins which combine with any dissolved haemoglobin entering the plasma as a result of the lysis of red cells. This prevents the haemoglobin from being excreted by the kidney, but the share of the haptoglobins in conserving the body's supply of iron is not fully understood. The literature of the biochemistry and genetics of the haptoglobins is extensive and complicated (see Giblett 1969).

Smithies *et al.* (1962) showed that the complex patterns found on starch-gel electrophoresis of serum or plasma could be attributed to three phenotypes, known as Hp 1-1, Hp 2-1, and Hp 2-2, which behaved genetically as though the proteins concerned were the products of two allelomorphic genes, *Hp¹* and *Hp²*; of the gene products, Hp¹ is more efficient than Hp² in removing free haemoglobin from circulating blood. In many populations, but especially among Africans, an absence or a very low level of haptoglobin is found in some individuals. This is in part the result of depletion of haptoglobin through haemolytic anaemia, but a genetical factor is also involved, apparently a variant of the *Hp²* gene, recognizable however only in the heterozygote known as Hp 2-1 M. Despite much investigation, the relation between genetics and physiology in this system is by no means fully understood.

The transferrins

The transferrins, or siderophilins, are proteins containing 5·5 per cent of carbohydrate, which combine with inorganic iron in the plasma and transfer it to the bone-marrow and other storage organs.

As shown by starch-gel electrophoresis, there is one common gene *Tf^C* and large numbers of variants, all of which are more-or-less rare and most of them extremely so. A great many population tests have been done with this system, not so much because of its intrinsic interest as because it is so easy to record the phenotypes when the more interesting haptoglobin system is being studied.

The variants of pseudocholinesterase

Red cells contain a cholinesterase which breaks down the acetylcholine produced at nerve endings, especially those controlling voluntary muscles: it thus prevents muscles from going into permanent spasm. Another cholinesterase, now usually called pseudocholinesterase, is present in the plasma;

its normal function is uncertain, but the existence of genetically determined variations in its activity came to light as a
result of using another acylated choline compound, succinyl
choline, as a muscle relaxant in surgical anaesthesia. This
substance blocks the action of acetylcholine on voluntary
muscles and so produces relaxation. Its effect is usually
moderated by, and after some minutes most of it is broken
down by, the action of pseudocholinesterase, so that the tone
of muscles, including those of respiration, returns to normal.
From time to time, however, individuals have been found
who remain relaxed and without spontaneous respiration
for prolonged periods. This has been found to be due to a
deficiency of plasmatic pseudocholinesterase activity. In
1953 Forbat et al. showed the condition to be familial. These
and other workers further investigated the condition and
showed the existence of a series of alleles, several giving weak
enzyme activity being distinguishable from one another
by their interaction with a variety of inhibitors including
fluorides.

The common gene giving rise to a normal enzyme is known
as E_1^u. The most frequently encountered gene with a product
of reduced activity is E_1^a, with a frequency of 1 to 2 per cent
in European populations. The gene producing a relatively
inactive but fluoride-resistant enzyme is E_1^f, the frequency of
which, where ascertained, is usually about 0·5 per cent in
Europe.

Some of the highest frequencies of the E_1^a gene are found
in oriental Jews, 7·5 per cent in those of Iran, 4·7 per cent in
those of Iraq, and 3·6 per cent in Yemenite Jews.

Besides the immediate products of the genes at the E_1
locus, some sera contain another similar esterase which is
partly independent genetically, known as C_5. It is most
easily demonstrated by electrophoresis in a rather acid gel,
when all the bands directly due to E_1 alleles come together,
and that due to C_5 moves slightly more slowly towards the
positive pole. To a first approximation this enzyme behaves
as the product of a gene at a locus independent of E_1, and
known as E_2, with dominant expression. However, it appears
not always to be expressed in the heterozygote, and the
expression is also affected by the E_1 genotype. This is, therefore, not a very useful genetic marker, but it is mentioned
here since a number of series of frequency determinations
have been published, including one series of Jews.

The immunoglobulin groups

The immunoglobulins are the antibodies of the pathologists
and immunologists, looked at from the standpoint of
structural chemistry. Each immunoglobulin molecule may
be regarded as consisting of two portions, a variable one and
a so-called 'constant' one. The variable part is structually
tailored to enable the molecule to act as an antibody to one
of the very wide range of pathogenic organisms and other
substances in the environment. Thus each individual is
potentially provided with an enormous variety of immunoglobulins. The 'constant' portion, on the other hand, carries
antigenic structures which are hereditary and peculiar to
each individual. An adequate account of the structure and
genetics of the immunoglobulins would need a book in itself.
Since we have found only one record of a Jewish population
tested for the Gm and Inv antigens, any long account would
not be justified here. It must suffice to say that the antigens of
the immunoglobulins are controlled by two non-linked
genetic loci, determining the Gm and Inv groups respectively.
The Gm locus is extremely complex, apparently consisting of
a considerable number of closely-linked sub-loci, rather like

that already described for the Rh groups. However, the number of separate antigens which can be produced is very much
larger than for Rh, and the number of different combinations of these which the complete chromosome segment can
determine is very large indeed.

As with the Rh system there has been much controversy
regarding notation, but basically each of the separate
antigens is assigned a number or a letter, and the complete
gene complex is symbolized by a considerable string of
numbers or letters.

Despite the complexity of the genetics, and the laborious-
ness of the laboratory tests for detecting the antigens, the Gm
system has proved highly informative in the study of populations and could potentially tell us a great deal about the
affiliations of Jewish populations. It is hoped that such
investigations will one day be undertaken.

The Inv system is similar to the Gm, but very much
simpler. Those wishing for further information on these
systems are referred to the forthcoming book by A. G.
Steinberg or to that by Giblett (1969).

The Gc groups

The Gc (or group-specific component) groups of the plasma,
originally discovered by Hirschfeld (1959), are the expression
of genetically controlled variants of one of the α-globulins.
They are usually detected and distinguished by immuno-
electrophoresis. Their importance, other than as genetical
markers, has recently been realized as a result of the discovery
by Schanfield et al. (1975) that the Gc proteins are vita-
min-D carriers in the plasma. They are the products of a
system of allelic genes, of which two, Gc^1 and Gc^2, are
relatively common. A considerable number of others are
known, all of which are everywhere rare, apart from Gc^{Ab}
found in the indigenous populations of Australia and the
island of New Guinea. As shown by Mourant et al. (1976b)
Gc^1 has in general, though with several exceptions, a high
frequency in sunny climates and Gc^2 a relatively high one in
dull ones, suggesting that the Gc types are affected by natural
selection related to the availability of vitamin D.

The Ag beta-lipoprotein system

There are several genetically distinct systems determining
sets of variants of human plasma lipoproteins, the two most
fully investigated being the Ag and Lp systems.

The detection of variants in all these systems depends
upon their antigenic properties, and tests are carried out by
the Ouchterlony method of gel diffusion, using human or
animal antisera.

The original anti-Ag precipitating antibody was found by
Allison and Blumberg (1961) in the serum of a multiply-
transfused patient. A number of other antigens have since
been found, all being controlled by a single but complex
genetic system of a similar type to those responsible for the
Gm serum groups and for the Rh blood groups.

The original antiserum, now usually referred to as CDB,
was at first regarded as specific for a single antigen, Ag(a),
but this and other sera used in the earlier work are now
regarded as containing mixtures of antibodies, a situation
which has led to much confusion in working out the genetics
of the system and distinguishing it from the other lipoprotein
systems.

Only a few Jewish populations have been tested for this
system, some with the original CDB serum, and some with
pure anti-Ag^x, but none are known to have been tested with
more than one antiserum—i.e., for more than one allele (or
group of alleles in the case of mixed reagent sera).

The system is of considerable interest but, in view of the paucity of data on Jewish populations, it is unnecessary to discuss it in detail here. A full account, with references, and relating particularly to population studies, is given by Mourant *et al.* (1976*a*).

The Lp beta-lipoprotein system

Tests for the antigens of the Lp system depend upon the use of immune rabbit sera in the Ouchterlony precipitin test. The system comprises a number of antigens, but the precise genetical status of most of these is not clear. Most tests, including all of those on Jews, have been for the Lp(a) antigen only. This antigen is the product of the *Lp^a* gene, which expresses itself both in the homozygote and the heterozygote. So long as tests are limited to those for this one allele, the total frequency of all the other alleles can be regarded as the square root of the frequency of negatives, and the frequency of *Lp^a* determined by difference. The more complex aspects of the system are farther discussed by Mourant *et al.* (1976*a*).

THE RED-CELL ENZYMES

Inside the boundary membrane of the red cell is a solution, possibly with an elaborate microscopic structure, containing haemoglobin and a variety of other substances, especially enzymes. Haemoglobin and a great many of the enzymes show genetical polymorphism, the initial detection of which is in most cases made by means of differing speeds of migration during gel electrophoresis. The enzymes have, of course, precisely-known functions and in some cases the products of allelic genes differ quantitatively in functional activity. For each of a considerable number of the enzymes there is at least one allele conferring low or absent activity, and homozygotes for such alleles may suffer from distinct congenital diseases.

The biochemical and biological background of the red-cell enzymes, and hence of their genetical variants, is of great complication and must ultimately be considered in interpreting the distribution of these variants in differing populations. However, in this book they are being treated mainly as genetical markers and in this chapter only a greatly simplified account of each system can be given. As already stated, a full account of them is given by Harris (1970).

The acid phosphatases

The acid phosphatase of red cells was first shown to be genetically polymorphic by Hopkinson *et al.* (1963) by means of starch-gel electrophoresis and appropriate staining. The precise function of this enzyme in the red cell is unknown.

Three principal alleles P^a, P^b, and P^c are found in many populations, and a fourth, P^r, is not uncommon in some African peoples, especially the Khoisan. The possible interaction of the acid phosphatases with other genetically determined characters in determining susceptibility to malaria has been described by Bottini *et al.* (1971) and Palmarino *et al.* (1975).

Glucose-6-phosphate dehydrogenase

It has long been known that some persons in many Mediterranean countries suffer from favism, a haemolytic anaemia precipitated by consumption of the common horse bean or broad bean, *Vicia faba*. During the Second World War it was found that some American Negro soldiers suffered from haemolysis when treated with certain anti-malarial drugs derived from quinoline, such as primaquine. Investigations by numerous workers finally traced both these conditions,

and a number of others such as jaundice among new-born Chinese infants in Singapore, to a deficiency, in the red cells, of the enzyme, glucose-6-phosphate dehydrogenase (G6PD). The complicated history of research on the genetics of this enzyme has been reviewed by many authors including Giblett (1969) who gives full references to previous work.

The enzyme (present also in other tissues besides red cells) is an essential catalyst in one of the body's methods of oxidizing glucose, known as the pentose-phosphate pathway, since it involves a breakdown from the 6-carbon molecule of glucose to a 5-carbon chain. One of the effects of a deficiency of the enzyme is a tendency of the red cells to haemolyse, especially, as indicated above, in the presence of certain drugs and other substances.

A very large number of variants of G6PD are now known but, in addition to the normal type, only three variants are generally recognized as being of any importance in population studies. The common type in all populations is known as B+, the B referring to its rate of migration on electrophoresis and the + to its normal enzymic activity. The common abnormal type in the Mediterranean area is B−, migrating at the same speed as the normal but with an activity between zero and 7 per cent of normal. It is also labile on heating. It is of course possible to determine the speed of migration by the standard electrophoretic method only if at least a trace of enzyme activity is present in the specimen when it reaches the laboratory.

In African populations two abnormal types are known, A+ which has normal activity but which migrates faster than the normal, and A− which migrates at the same speed as A+ but has only 8 to 20 per cent of normal activity.

The variants of G6PD are determined by a series of allelomorphic genes on the X chromosome. Thus a male with only one X chromosome can have only one type of enzyme, but a female, having a pair of X chromosomes, can be heterozygous and have genes for two different alleles. It follows that in males the gene frequencies are identical with the phenotype frequencies.

Tests are of three main kinds, screening tests for the presence of enzyme deficiency, precise quantitative tests to measure the degree of deficiency, mainly of use in identifying rare variants, and tests by starch-gel electrophoresis. Provided that blood specimens are fresh, so that all the variants retain some enzyme activity, electrophoretic tests will, in males, and to a large extent also in females, show both the relative electric charge and the activity of any variants present.

Glucose-6-phosphate dehydrogenase deficiency, especially that of the Mediterranean B− type, can, particularly through favism, lead to serious illness with an appreciable mortality. The genetical problem which it presents is thus similar to that of the abnormal haemoglobins and the thalassaemias; we need to explain how a gene with deleterious effects can nevertheless maintain very high frequencies in certain populations. In certain areas, and especially in the Mediterranean region, the distribution of the gene shows a considerable degree of correspondence with that of recent, and in some cases still present, malignant tertian malaria. There is also some direct evidence for protection of the individual against the effects of malarial infection (Allison and Clyde 1961; Harris and Gilles 1961; Bienzle *et al.* 1972).

A major part of the world distribution of the deficiency lies in a belt extending from the Mediterranean area through south-west Asia and India to south-east Asia including Indonesia. Throughout this belt it appears to be mainly of the B− electrophoretic type. The highest frequencies of this, or any, type of deficiency yet recorded are in Kurdish Jews of Iraq. Only slightly lower are the frequencies in Kurdish Jews of Iran, and in Arabs of the eastern oases of Saudi Arabia (Gelpi 1965, 1967). Rather high frequencies have also been reported in a number of other Iranian populations and in the Parsis of India whose ancestors came from Iran. The high and variable frequencies found in the Near East, and in many Jewish populations, present a problem of considerable interest which is further discussed on pp. 29 and 55.

The variants of 6-phosphogluconate dehydrogenase

The enzyme 6-phosphogluconate dehydrogenase takes part in the hexose-monophosphate shunt, that is to say, a relatively minor series of reactions branching off from the main aerobic glucose metabolic cycle, and converting hexoses into pentoses needed for making nucleic acids. It catalyses the next step in the chain of reactions after that catalysed by glucose-6-phosphate dehydrogenase, and causes the conversion of 6-phosphogluconate to ribulose-5-phosphate.

Numerous variants are known to exist, determined by a set of allelic genes.

In nearly all populations PGD^A has a frequency exceeding 90 per cent and PGD^C comes next in frequency. All other variants are rare and sporadic. The frequency of gene PGD^C varies between 1 and 4 per cent in European populations, and tends to be higher in the south. It is somewhat higher in the Near East and in Jewish populations; in Africa frequencies of this gene are mostly around 6 per cent but reach about 15 per cent in the Ethiopians, the Beja of the Sudan, and the South African Bantu. They are on the whole rather higher still in Asia, varying from 7 per cent in much of the Far East to 10 per cent in Nepal and 23 per cent in Bhutan. The gene is absent in most Amerindian populations, including the Caribs of Dominica who, however, have 1 per cent of PGD^R.

In Ethiopia (Harrison et al. 1969) two populations living at 1500 and 3000 metres respectively were compared for a wide range of genetic factors. The only system which showed significant differences was that of the 6-phosphogluconate dehydrogenase variants. The frequency of the gene PGD^C was found to be 5·3 per cent at 1500 m and 13·5 per cent at 3000 m. However, as tests for more than twenty systems were carried out, it could be expected that one of them would by chance alone show apparently significant differences between the populations at the 5 per cent level. Nevertheless, this result initially appeared to suggest an association between the frequency of the PGD^C gene and altitude. The high frequency in Bhutan seemed at first to support this view, but other high frequencies in Asia, in peoples living at lower levels, make any causal relationship appear much less likely. The possibility should nevertheless be borne in mind if further relevant data should become available.

Phosphoglucomutase

The enzyme phosphoglucomutase catalyses the reversible isomeric conversion of glucose-1-phosphate to glucose-6-phosphate.

Enzymes with this activity occur in various tissues but we are concerned here only with those found in red cells and their haemolysates. The pattern given by such haemolysates is at first sight bewildering in its complexity but, following the analysis of the genetical situation by Spencer et al. (1964) it is not difficult to understand in theory or, as a rule, to interpret in practice.

Spencer et al. showed that two non-linked autosomal loci, PGM_1 and PGM_2 are involved. Each gene product of PGM_1 gives rise to two bands and each product of PGM_2 to two main bands and a faint faster one. Heterozygote patterns are simply those of the gene products concerned, without additional bands due to hybrid molecules. Most of the bands due to PGM_2 alleles move faster towards the positive pole than most of those due to PGM_1 so that in all the more common genotypes there is no confusion between them, but in some of the rarer genotypes there is overlap, leading to difficulty in identification.

In population studies the most useful system is PGM_1 with two common genes PGM_1^1 and PGM_1^2 varying widely in frequency in different populations, and at least six rarer genes. With a very few known exceptions, the frequency of PGM_1^1 always exceeds 50 per cent. It is therefore convenient to describe populations in terms of their frequencies of the gene PGM_1^2, and of the rarer genes where they occur.

In the present context it will suffice to consider distributions only in Europe, Africa, and Asia apart from the southeastern islands. European populations have frequencies of PGM_1^2 increasing from 14 per cent in Ireland and 18 per cent in Iceland, eastwards and southwards to 31 per cent in Greece and 32 per cent in Turkey. Lapp populations mostly have higher frequencies, up to 53 per cent. Near Eastern and Jewish populations mostly have 25 to 30 per cent but the Habbanite (Arabian) and Cochin (Indian) Jews have respectively 58 and 59 per cent.

Frequencies of PGM_1^2 in India exceed 30 per cent but the Chinese and Japanese have between 20 and 25 per cent, the former apparently having also about 1 per cent each of PGM_1^6 and PGM_1^7 (but here the possibility of variants at the PGM_2 locus needs to be considered).

African Negroid populations vary between 15 per cent of PGM_1^2 in the Babinga Pygmies of the Congo and 24 per cent in the Yoruba of Nigeria and the Hadza of Tanzania. The more Caucasoid Amhara of Ethiopia have just under 30 per cent.

Most populations tested, other than Africans, lack variants at the PGM_2 locus, being uniformly homozygous for PGM_2^1, but most Negroid populations have 1 to 3 per cent of the PGM_2^2 gene.

Adenylate kinase

The enzyme adenylate kinase reversibly catalyses the reaction:

adenosine diphosphate $\leftrightharpoons$ adenosine triphosphate
$+$ adenosine monophosphate

$$(2ADP \leftrightharpoons ATP + AMP)$$

Numerous phenotypes are known, which are the products of a series of allelic genes, but the only ones of any substantial importance in population studies are AK^1 and AK^2, and even the latter seldom exceeds 5 per cent in frequency. In European populations the gene AK^2 nearly always has a frequency between 2·5 and 6 per cent but this falls to about 1 per cent in Lapps and Sardinians. In the populations of the Near East, including Jews, frequencies are similar to those found in Europe. In Africa AK^2 levels are about 1 per cent and several Negroid populations tested have shown only the AK^1 allele. The Amhara of Ethiopia and the Beja of the Sudan, though apparently possessing about 50 per cent of Caucasoid genes, have respectively only 1·3 and 0·5 per cent of AK^2. The highest known frequencies of AK^2 are in Indian and Pakistan subjects tested in England, with 10 and

13 per cent respectively, and 7 per cent in Iranian Kurds, but in Thais and Chinese the level falls to 0·2 per cent.

With the large variations found in the continent of Asia the system may ultimately prove of considerable value in the study of Oriental Jewish populations.

Malate dehydrogenase

Malate dehydrogenase catalyses the reversible oxidation of malate to oxalo-acetate. It exists in two forms, one present in solution in cytoplasm, including red cells, and one localized within the mitochondria. These are the products of genes at separate autosomal loci, not closely linked. We are concerned here solely with the soluble variety. Nearly all populations, including all the Jews so far tested, are uniformly homozygous for one common gene; variants are known in a few populations but are everywhere exceedingly rare except in Papua and New Guinea.

The red-cell peptidases

The red-cell peptidases are enzymes which catalyse somewhat selectively the hydrolysis of certain di- and tri-peptides. They are the products of genes at five loci. When red-cell haemolysates or sonicates are subjected to starch-gel electrophoresis and then stained for peptidase activity the products of genes at any one locus are seen to fall into a comparatively narrow band. As shown by Lewis and Harris (1967), five zones can be distinguished, which are lettered A, B, C, D, and E, migrating at increasing speeds towards the anode. There is some overlapping between these zones, but the products of genes at any one locus can be distinguished from all the others by certain reactions. For instance, only the variants of Peptidase A lyse the valyl-leucine linkage and only those of Peptidase B lyse leucylglycyl-glycine. Peptidases A, B, C, and D have been found to exhibit genetic variation, but Jews have so far been tested only for peptidase A variants. Europeans (including a group of Jews from Germany) and Indians are nearly 100 per cent homozygous for $Pep\ A^1$, but Negroes have approximately 10 per cent of the $Pep\ A^2$ gene.

The variants of adenosine deaminase

The enzyme adenosine deaminase is an aminhydrolase which catalyses the deamination of adenosine to inosine. Genetically determined variants were first demonstrated by means of starch-gel electrophoresis by Spencer, Hopkinson, and Harris (1968). These were shown to be the products of a pair of allelic genes ADA^1 and ADA^2 at an autosomal locus. Other genes, ADA^3, ADA^4, and ADA^5, are so rare as to be of very little interest in populations so far studied, and even ADA^2 is not known to exceed 20 per cent in frequency in any population.

The few European data which we possess seem to show an eastward increase in the frequency of ADA^2 from 5 per cent in Ireland and 6 per cent in England to 8 per cent in Germany and 9 per cent in Italy. The same trend continues in Asia with 12 per cent in Iranian Kurdistan, India, and Nepal. In Papua 12 per cent of the gene is found again in Kar Kar Island but 17 per cent in Goroka on the mainland. In Africa the average frequency is under 1 per cent.

Glutamic-pyruvic transaminase variants

Glutamic-pyruvic transaminase catalyses the reversible conversion of L-alanine and α-ketoglutarate to L-glutamate and pyruvate. Like certain other enzymes it exists in two molecular forms, one present in mitochondria and another (soluble) present in cytoplasm. In mature red cells only the soluble form is present and by electrophoresis of red-cell haemolysates Chen and Giblett (1971) have shown it to be genetically polymorphic, with three phenotypes representing the homozygous and heterozygous expression of two autosomal allelmorphic genes, Gpt^1 and Gpt^2. Chen et al. (1972) and several other authors have carried out surveys which show the frequencies of these genes to vary widely in different populations.

Frequencies of the Gpt^2 gene range from 13 to 69 per cent, averaging about 18 per cent in Africans and 46 per cent in Europeans. The system thus promises to be a most useful one in population studies. Jewish populations show frequencies which mostly lie at the lower end of the European range.

THE HAEMOGLOBINS

Relatively little work has been done on the incidence of haemoglobin variants in Jewish populations, but the importance of the haemoglobins in the story of genetic factors in resistance to malaria makes it necessary to include here a brief account of them. For fuller information the general scientific reader is referred to the work of Lehmann and Huntsman (1974) while for details on the distribution of the variants in different populations reference should be made to the monograph of Livingstone (1967).

The haemoglobin molecule consists of four polypeptide chains, two identical α chains, and two identical β chains, each chain having attached to it one heme group, consisting of an iron atom surrounded by a porphyrin ring which itself contains four pyrrole rings. It is the iron which combines reversibly with oxygen to give haemoglobin its oxygen-carrying power, but the surrounding chains of porphyrin and protein determine the precise conditions under which the iron takes up and gives off oxygen.

The amino-acid sequences in the α and β chains are controlled by genes at separate loci which are not detectably linked, but families yielding potential linkage data are inevitably very rare. The great majority of people everywhere have one type of haemoglobin, normal adult haemoglobin, or Haemoglobin A. About one hundred abnormal types are known, of which the majority have abnormal β chains. Most of the abnormal types are extremely rare, and only four of them, Haemoglobins C, D, E, and S have frequencies which are sufficiently high to be of interest in population studies. These are all the results of substitutions in the β chain of normal adult haemoglobin.

Haemoglobin S

Of the abnormal haemoglobins Haemoglobin S or sickle-cell haemoglobin was the first to be discovered, in 1949 by Pauling et al., who demonstrated that an abnormal haemoglobin was present in the blood of persons showing the sickling phenomenon.

Persons who are homozygous for the Haemoglobin S gene have all their normal adult haemoglobin replaced by the abnormal type. This haemoglobin in the deoxygenated or reduced state has a very low solubility in the internal fluid of the red cell, and is therefore precipitated in the form of angular crystals (strictly, liquid crystals or tactoids) which distort the cells into an angular or sickle shape in which they become unduly susceptible to mechanical damage and destruction. Most homozygotes die in infancy from sickle-cell anaemia. Heterozygotes, with one gene for the β chain of Haemoglobin S and one for that of Haemoglobin A, have both types of haemoglobin in their red cells, though more of

A than of S. They suffer only very slight disabilities, but it has been shown by Allison (1954) and many subsequent workers that they are much more resistant to the harmful effects of malignant tertian malaria (that due to *Plasmodium falciparum*) than are 'normal' Haemoglobin A homozygotes. It is clear from numerous population surveys that by natural selection in recent centuries the frequency of the *HbS* gene has become raised in various African populations in response to the varying degrees of endemicity of this form of malaria. Though *HbS* homozygotes have died of sickle-cell anaemia, *HbA* homozygotes have, as just mentioned, had a high mortality from malaria and the favoured heterozygotes have survived selectively to maintain the frequency of both genes in the next generation.

THE THALASSAEMIAS

The thalassaemias are pathological conditions, but they are due to abnormal single genes, and they are sufficiently frequent in some populations, including some Jewish ones, to have considerable significance in population studies. There are two main types of thalassaemia, α-thalassaemia and β-thalassaemia, due respectively to genes which suppress the production of α and β chains of haemoglobin.

To enable some of their manifestations to be understood, it is necessary to mention yet another type of haemoglobin, present in small amounts in the red cells of normal adults. This is Haemoglobin A$_2$, with a constitution $\alpha_2\delta_2$. The molecule thus contains two normal α polypeptide chains and two chains which are closely similar to, but not identical with, β chains, and are controlled by separate genes. Thus when there is interference with the synthesis of β chains that of δ chains is unaffected, and Haemoglobin A$_2$ is present in relatively increased amount. However, anything which suppresses the production of α chains will limit the production both of Haemoglobin A and of Haemoglobin A$_2$. The relative amount of Haemoglobin A$_2$, as seen on electrophoresis, is thus an important criterion, not only in detecting β-thalassaemia, in population screening tests, but also, if thalassaemia is known to be present, in distinguishing one type from the other. The production of foetal haemoglobin also is normal or sometimes greatly increased in β-thalassaemia, but is reduced in α-thalassaemia. In both the main varieties of thalassaemia the abnormal gene appears to interfere with the transcription, into an amino-acid sequence, of a gene with normal DNA constitution coding for an α or a β chain respectively.

Beta-thalassaemia has been much more fully studied than the alpha variety, and is better understood, though it is only in recent years that the distinction between the two varieties (each with subtypes) has become clear.

Heterozygotes for β-thalassaemia (possessing one normal gene for the critical transcription stage and one abnormal one) suffer from a mild anaemia, while homozygotes have a severe anaemia and few of them, without intensive medical care, survive childhood. Death does not take place anything like as early as with sickle-cell anaemia, but the formal genetical situation is almost the same in the two diseases. It is also almost certain that the frequency of β-thalassaemia is maintained through resistance of the heterozygotes to malaria, by a mechanism analogous to that just described for Haemoglobin S. This was in fact suggested by Haldane in 1949. The evidence, however, comes entirely from population surveys; resistance of thalassaemic individuals to malaria has not so far been demonstrated.

THE HISTOCOMPATIBILITY ANTIGENS

Our knowledge of the use of the histocompatibility antigens in population studies, and of their remarkable disease associations, is of very recent but extremely rapid growth and little about them is therefore to be found in any of the books listed earlier, dealing with other hereditary blood factors. A comprehensive account of these antigens, and especially of their genetics, is given by Svejgaard *et al.* (1975), and a description of their distribution in the populations of the world is provided by Dausset and Colombani (1973). Their disease associations are dealt with by Ryder and Svejgaard (1976) and by Dausset and Svejgaard (1977).

These antigens have been known for many years as being present on leucocytes and platelets, but they have now been shown to be the principal antigens responsible for compatibility and incompatibility of tissue and organ grafts in man. This realization of their clinical importance has led to much research on their genetics, more recently to work on their distribution in different populations and, more recently still, to investigation of their remarkable associations with diseases.

The antigens, though present in most tissues of the body, are usually identified by tests on lymphocytes. They are dependent on a set of closely linked genes on a short segment of a chromosome (No. 6), an arrangement similar to that postulated in Fisher's well-known model of the Rh blood group genes, but more complex. Four closely-linked loci are involved, now named in serial order A, C, B, and D. Each locus may be occupied by any one of a large range of polymorphic genes, each gene producing a single antigen (but some genes are silent, which probably means that they are producing antigens for which diagnostic antibodies have not yet been found). The combination of four genes on a given chromosome is known as the haplotype. Because of the great range of possibilities at each locus, the number of possible haplotypes is very large, and that of genotypes (combinations of two haplotypes) almost astronomical. It is this complexity which gives the system its great potentiality in differentiating populations. The study of the close association of particular genes and antigens with specific diseases has hardly begun, but in this respect the system promises to be the most important and exciting of any yet found in man.

POLYMORPHISMS NOT ASSOCIATED WITH THE BLOOD

A number of human biochemical polymorphisms are known which express themselves other than in some component of the blood. Two of these must be considered here, since they have been the subject of investigations on Jewish populations: the phenylthiocarbamide tasting system and the acetylator system. Lactase deficiencies also demand consideration.

The phenylthiocarbamide tasting system

There is no recent comprehensive work on the tasting system, but a history of research with a comprehensive bibliography is given by Mourant *et al.* (1978). In 1931 Fox observed that to some individuals the simple chemical compound, phenylthiocarbamide, has an intensely bitter taste, while to others it appears tasteless. The difference between individuals is not in practice an absolute one, but by determining for each person a threshold concentration of phenylthiocarbamide in solution, which is just tastable, it is possible to classify nearly every one as either a taster or a non-taster. There are two allelomorphic genes, for tasting and non-tasting, the former being dominant in expression, so that heterozygotes are tasters. Tests have been carried out in various ways. The only fully reliable ones depend on determining the taste threshold with a set of solutions of different dilutions. Single-solution tests are less reliable, and those with crystals or dried solutions on filter paper still less so. For this reason Table 31 shows the methods of testing where these are known.

Populations show a wide range of gene frequencies but, unfortunately for their use as taxonomic markers, a very large number of populations, including many of the Jewish ones tested, fall into a class with little variation from 50 per cent of each gene.

The chief interest of the system lies in its physiological and pathological implications. The taster–non-taster phenomenon applies to a considerable number of substances chemically related to phenylthiocarbamide, and these substances are all suppressors of thyroid function, some being used therapeutically for this purpose, and others occurring naturally in some food plants. A knowledge of this thyroid-suppressing action has led to the finding that persons with nodular non-toxic goitres show an excess of non-tasters (Harris et al. 1949), while patients with diffuse toxic goitres include an excess of tasters (Kitchin et al. 1959). There is moreover some evidence for an association of tasting with the level of thyroid function in persons without goitre, and even with such phenomena as the degree of bone development at puberty in relation to age.

The acetylator system

There is no readily accessible comprehensive account of the acetylator system but Mourant et al. (1976a) have given an account and a key bibliography of it.

Harris et al. (1958) and Evans et al. (1960) showed that in certain individuals the antituberculosis drug isoniazid disappeared more quickly from the blood than in others. The rate of disappearance with the earlier tests appeared bimodal, with the types genetically determined and fast metabolism expressing itself as the dominant condition. More sophisticated tests enable the heterozygotes to be distinguished almost, if not quite, completely from both the homozygous types. The metabolism in question is the acetylation of the drug, and the same system controls the acetylation of a number of other substances including the drug dapsone used for the treatment of leprosy. The system is thus of some importance in clinical medicine. The frequency of rapid acetylators varies over a wide range, being especially high in Mongoloid peoples. So far, only a few Jewish populations have been tested.

Primary adult lactase deficiency

It is remarkable that adult lactase deficiency, a common condition with marked clinical manifestations, was not clearly recognized until 1965 (Haemmerli et al. 1965). Even in non-Jewish Europeans it has a frequency near ten per cent, and is liable to give rise to marked gastrointestinal disturbances following a meal containing large quantities of milk.

In man as well as in other mammals the newborn offspring require a supply of lactase in the intestine to enable them to metabolize the lactose of the mother's milk. Infantile lactase deficiency in man is a very rare and serious disorder which has long been known.

Until the invention of dairy husbandry, adults, and infants after weaning, did not need lactase, and it is now known that in most populations the majority of modern adults do not possess it. However, in the majority of northern Europeans, and in a substantial proportion of persons of other races, lactase is present and enables those with it to enjoy a diet containing much milk. The presence or absence of the enzyme in the individual adult appears to be inborn, and not to be conditioned, like some enzyme deficiencies, by the presence or absence of the substrate (lactose) in the diet.

Genetic studies and population surveys are hindered by the limitations of the available methods of testing. The most satisfactory method is a quantitative estimation of lactase in biopsy specimens of the jeunal mucosa, but this can be done only on surgical patients and small numbers of dedicated volunteers.

Lactose tolerance tests depend upon the measurement of the blood glucose level after ingestion of a standard quantity of lactose, and can readily be performed on large numbers of persons, but the interpretation of results is at present somewhat arbitrary.

It does not appear to be known whether or not the normal lactase of infants has the same chemical structure and is the product of the same gene as that of adults. This could be ascertained without great difficulty by standard protein characterization methods.

Almost certainly infantile lactase deficiency is a recessive condition due to homozygosity for a deficiency gene. The few data which we possess (mainly from Welsh et al. 1968) on the genetics of the adult deficiency appear to be consistent with this condition also being a recessive manifestation. Ferguson and Maxwell (1967) have suggested a system of three allelic genes: L, the normal with dominant expression, l_1 for which deficient infants are homozygous, and l_2, homozygous in deficient adults. They further suggest that 'the genotype $l_1 l_2$ could result in the rare situation of hypolactasia presenting in childhood, with normal milk tolerance in infancy'. Another possibility, if normal infant and adult lactases are the products of the same gene, is that adult deficiency is caused by a suppressor gene, not closely linked to that for the enzyme itself.

In view of the great importance of adult lactose intolerance, in relation to the nutrition of individuals and communities, it is most desirable that more genetic studies should be carried out by precise methods and on adequate numbers of persons. Because of the present uncertainty about the genetics, we have in Table 33 shown only phenotypes, and made no attempt to compute gene frequencies. It must, moreover, be realized that even the definition of phenotypes is somewhat arbitrary, depending, among other things, upon the level of lactase activity which is chosen as the lower limit of 'normality'.

Among Europeans, the British include 22 per cent of deficients, the Swiss, 6 per cent on a sample of only 17, and U.S. 'Caucasians', 13 per cent, but Greek Cypriots have 88 per cent, and Palestinian Arabs 81 per cent, suggesting the possibility of a high frequency in much of the Mediterranean area. Negroes, Asiatic and American Mongoloids, and Australian Aborigines have frequencies from 67 per cent to 100 per cent. Frequencies in Jewish populations are discussed on p. 54.

3

THE JEWS IN PALESTINE

THIS chapter is entitled, the Jews 'in', not 'of' Palestine for, apart from the Samaritans who have a chapter to themselves, no community of Jews to be found at present in Palestine has lived continuously in that country since biblical times. There is now in process of formation a new composite community of Jews born in Palestine, which deserves the closest biological study, but the pursuit of its development is a task for future investigators.

Here we are concerned with the history and biology of the former population of Palestine, as it affects the genetics of the Jews of the Dispersion.

Palestine, a key area on the road between Asia and Africa, has seen great movements of peoples, and has been under a confusing succession of jurisdictions. This is not a history book and no attempt has been made to go back to primary sources other than the Bible; a great many specialized works listed in the bibliography have been consulted but use has been made chiefly of the work of Hitti (1961) in trying to trace the continuing thread of events.

We know much more about the language than about the physical anthropology of the tribes and nations that have passed through Palestine. Language is a cultural character-istic, and peoples have often changed their language in a short space of time while maintaining their genetical identity, but language can nevertheless provide important clues to the ancestry and biological relationships of peoples.

FROM MOSES TO SOLOMON

Few historians would accept the historical parts of the Pentateuch as a precise record, but they may be regarded as embodying genuine oral traditions of the origin and wander-ings of the Children of Israel. Apart from the sojourn in Egypt, the Israelites moved in a region of Semitic-speaking peoples. They almost certainly understood one anothers' languages and probably shared a common ancestry.

Abraham, head of a tribe of nomadic herdsmen, is said to have come from Haran in Mesopotamia. After spending a few generations in Palestine the tribe was driven by famine to Egypt where the Semitic-speaking Hyksos or Shepherd Kings were in power, their capital Avaris being at the north-east corner of the Delta, almost on the borders of Palestine. The Hyksos were expelled by Pharaoh Ahmose (1580–45 B.C.) and it must be supposed that the Hebrews were at this time reduced to slavery. They escaped at the Exodus, about 1225 B.C., under Moses who was probably a rebel Egyptian, and after many years of wandering in the wilderness, and adopting Jahweh as their sole God, they crossed the Jordan at Jericho and began settling in Palestine. The biblical story is made to apply to all the Israelite tribes, but it is probable that not all these entered the country at the same time. Also they were all closely akin in language and probably in ancestry and traditions to the Canaanites who were already there. Some of their penetration may have been peaceful but cities especially had to be besieged and stormed. Partly con-

quered, partly absorbed, the Canaanites ultimately disap-peared as a distinct people.

However, the most serious opponents of the Israelites were the Philistines of the southern part of the coastal plain. They also were newcomers, probably refugees from the Greek Islands and Asia Minor, fleeing before the classic mainland Greeks. The second King of the Israelites, David, finally unified almost the whole of Palestine and set up his capital in Jerusalem where his son Solomon built the first Temple. David ascended the throne about 1004 B.C.

At this stage we may visualize Palestine as being inhabited by a more-or-less uniform Semitic-speaking nation. To the south-east the Philistines were causing relatively little trouble. To the north the Phoenicians were a friendly Semitic-speaking people concerned mainly with manufac-tures and maritime commerce, sending their ships to all parts of the Mediterranean where they set up colonies (Carthage was founded about 814 B.C.). Some authorities think that Israelites accompanied these colonizing expeditions and so began the great dispersion.

To the south and east the desert peoples were at peace and, farther away, the great warlike nations, the Hittites to the north, Assyria and Babylonia to the east, and Egypt to the south-west, were biding their time. The Israelites themselves were concerned largely with the building of the Temple.

When the Israelites entered Palestine they spoke Aramaean, but they subsequently came to speak another Semitic lan-guage which became the classical Hebrew of the Bible. The latter differs only slightly from Phoenician, and it appears that the Israelites, hitherto illiterate, adopted that language, with slight variants, together with the alphabet in which it was written. About 500 B.C. the everyday language had changed to another variety of Semitic, Aramaic, but Hebrew remained the sacred language and was recently revived as the common language of the State of Israel.

THE FIRST DISPERSION

After the death of Solomon, about 923 B.C., petty tribal and nationalistic squabbles led to separate monarchies being set up for Judah with its capital in Jerusalem, and for Israel, composed of the ten northern tribes, with the capital at Shechem (corresponding approximately with the modern village of al-Balatah, and near the newer town of Nablus). Later the town of Samaria became the capital.

Quarrels between and within the two kingdoms continued, and even war between them, giving opportunities to the great eastern powers of Assyria and Babylonia.

Finally in 732 B.C. Tiglath-Pileser, king of Assyria, cap-tured Damascus and deported many of the inhabitants, and in 722 his successor, Shalmaneser V, captured the city of Samaria. It was however his usurping successor, Sargon II, who reaped the fruits of this victory, taking away many of the inhabitants of Samaria to 'Halath, Hathor and the cities of the Medes', and replacing them by aliens from other subject territories. This was the real beginning of the Dispersion or Diaspora of the Israelites; the fate of the captives will be considered in later chapters.

Subsequently the depopulation, and dislocation of government, in the land of Samaria led to devastation by lions. This was interpreted as being due to a failure to give the correct service to Jahweh, the god of the land, and priests were therefore sent back to restore worship in the correct manner. This was the beginning of what was almost certainly a syncretic form of Jewish worship, which led to the later antagonism with Judaea at the time of the rebuilding of the Jerusalem temple, further described on pp. 18–19.

At the time of the capture of Samaria the Assyrians, with their capital at Nineveh, near modern Mosul, already controlled Babylonia, and later they were victorious over Egypt. Meanwhile the kingdom of Judah continued its rather precarious existence. The Chaldaeans or Neo-Babylonians now obtained control of Babylonia, and in alliance with the Medes of north-western Persia they marched against Nineveh and destroyed it completely. The Neo-Babylonians under Nebuchadnezzar II then defeated Egypt and in 586 B.C. took and destroyed Jerusalem and the Temple. They took into captivity in Babylon many of the Jews, especially of the priestly and ruling classes. This was the beginning of the great Jewish colony in Babylon which, for over 2500 years, until A.D. 1948, remained one of the largest and for the greater part of the time one of the most influential centres of world Judaism. But it was also the beginning of the colony in Egypt, for many at this time fled thither, including the prophet Jeremiah.

The Persians were now asserting their power in the region and Cyrus, king of the Medes and Persians, having overcome the Greek king Croesus of Lydia on the west coast of Asia Minor, turned towards Babylon which, ruled by the pleasure-loving crown prince Belshazzar, fell in 538 B.C. Cyrus allowed exiled Jews to return to Jerusalem, and the Persian rulers continued to allow deported peoples to return to their homes. Jews went to Jerusalem under Nehemiah in 445 B.C. to repair its walls, and (probably some years later) under Ezra to rebuild the Temple.

Many of the Jews whose ancestors had been forced to settle in Babylon evidently enjoyed the life there, and, despite the rebuilding of Jerusalem, many seem to have remained voluntarily in Babylon, and even Nehemiah returned there after rebuilding the walls of Jerusalem.

The Persian empire extended from Egypt to the Indus, but in the middle of the fourth century B.C. it began to weaken until Alexander the Great, son of the Greek Philip of Macedon, swept it all away. Tyre fell to him in 332 and Gaza and then the Nile delta followed, where in 331 he founded Alexandria, later to figure largely in Jewish history. At the most westerly point on his marches he was welcomed as a Pharaoh by the priests of Amon (Greek Ammon) in the Siwa Oasis; he engrafted the worship of Zeus (Jupiter) on to that of Ammon, whence come the modern names of ammonites (fossil molluscs) and ammonia from sal ammoniac (distilled from camels' dung), both products of the oasis.

After conquering most of the known world he died at Babylon in 323, at the age of 33.

As so often in history, the realm of this mighty conqueror barely survived his death, though the world could never be the same again.

Alexander's empire was divided between four of his generals, of whom Seleucus, reigning from 312–280, took the greater part, including Asia Minor and Syria, which itself included Palestine, while Ptolemy took Egypt and founded a dynasty of Ptolemies.

Seleucus established a uniform calendar, beginning in 312 B.C., for the whole of his dominions, and it is due to him and his descendants and successors the Seleucids that the Greek language and culture became those of the east Mediterranean.

In 241 Ptolemy III of Egypt took the coast of Syria and Asia Minor from Seleucus II. The son of the latter, Antiochus the Great, won most of the area back but was defeated in 190 B.C. by newcomers, the Romans, at Magnesia in Asia Minor, and in 188 he was forced to cede most of Asia Minor to them. Again after a successful campaign against Ptolemy IV (who was captured) in Egypt he was forced by the Romans to raise the siege of Alexandria.

Though under centuries of Persian rule the Jews were a subject people, they do not seem to have been oppressed, nor were they so under the early Seleucids. Following the initial flight of many Jews to Egypt after the destruction of Jerusalem, it is likely that more moved there from time to time, and then the Ptolemies actively encouraged the development of the Jewish colony at Alexandria. But the policy of Hellenization followed by Antiochus the Great finally drove the Jews to revolt, when he proclaimed himself Theos Epiphanes (God manifest) and set up an altar to Zeus in the Temple at Jerusalem. Judas, a member of the priestly Hasmonean family, who was surnamed Maccabeus or the hammerer, together with his brothers, led the revolt and for nearly a century under the Maccabean dynasty the Jews enjoyed first religious and then also political freedom, until the arrival of the Romans.

In his triumphal progress through the Near East, the Roman general Pompey took Jerusalem in 63 B.C. For most of the subsequent century Palestine was ruled either directly by the Romans or indirectly through puppet kings of the Herodian family. The founder, Herod the Great, was an Idumaean who professed the Jewish religion, but maintained his authority by a succession of assassinations, the victims including members of his own family. He ruled Judaea (a province which included Samaria) from 37 to 4 B.C., the year 6 B.C. on the universal scale of dates being now widely accepted as that of the birth of Christ. From A.D. 6 to 41 Judea was under Roman Procurators, of whom Pontius Pilate held office from 26 to 36. After a short period under Herod Agrippa, Judaea again in A.D. 44 came under the rule of a succession of Procurators, and from about A.D. 60 relations between rulers and ruled became more and more tense. In 66 the priests stopped the sacrifices which had hitherto been made in the temple on behalf of the emperor. This was open revolt, to be put down by military action. The Roman campaign was led first by Vespasian and, when he became emperor in 69, by his son Titus (later emperor), who took Jerusalem in 70 with great loss of life among the defenders. The Temple, against Titus's orders, was burned down. The representation, on the Arch of Titus in Rome, of the triumphal march with the Temple treasures on display, is well known.

THE FINAL DISPERSION

This was the beginning of the final dispersion of the Jews from Palestine. For the next 1800 years Jerusalem was to be no longer a mainly Jewish city, though for some centuries large numbers of Jews were still to be found in Palestine. At this time, however, while Babylon and Alexandria were the main centres of Jewish religion and culture, we can see from the Acts of the Apostles and the book of Maccabees that there were Jewish communities in practically every country and large town in the east Mediterranean area, including Rome.

It was to these that the Jews of Palestine gravitated over the first few centuries of the Christian Era, to give rise in due course to the Jewish communities of Europe. For the Jews of Palestine political activity was completely suppressed but religious worship and thought continued.

There were at this time a number of sects or parties in Palestine with widely-differing attitudes to religion, to daily life, and to politics. The Essenes were extreme ascetics who retired to the desert to practise Judaism in what they regarded as its purest form. The Pharisees refrained from political activity but remained in the community as its moral and spiritual leaders, while the Saducees were much more directly concerned with politics and maintaining relations with the ruling power. The Zealots were political activists, devoted to the Torah, but prepared to undertake armed rebellion for the sake of their faith.

During the siege of Jerusalem by Vespasian and Titus, the Pharisee leader, Rabbi Jochanan ben Zakkai, had realized that the city was doomed and that the only hope for the future of Judaism was to continue solely as a religious movement, based on the Torah itself and the synagogues rather than on the Temple worship. He therefore went to Jabneh on the coast to establish a cultural centre for the Jewish people. He is said to have sought out the general Vespasian and secured from him the gift of the already existing Jabneh religious school. When the news of the destruction of Jerusalem arrived he 'rent his garments', but continued with the building up of the spiritual centre.

The authority of the Sanhedrin which he set up soon became recognized by Jews everywhere. One of its duties was to establish the calendar of religious fasts and feasts. The centre also promoted the *Midrash* or oral exposition of the Torah.

Next, in Jabneh, the Midrashim or commentaries began to be committed to writing, and it was here also that the final acceptance of the Song of Songs, Ecclesiastes, and Esther into the biblical canon took place, as well as the important editing of the whole Old Testament, and the preparation of a new Greek translation of the latter. Then, when the emperor Hadrian had a temple to Jupiter set up on the ruined site of the Temple, the Jews once again in A.D. 132 rose in revolt under Simon ben Kosebah. The revolt was brutally suppressed and followed by more severe repression of the Jews. Hadrian realized the importance of the Torah and the Jewish religion as a rallying force and attempted to abolish the latter completely.

The Jabneh school was abolished and many Jews died for their faith. In order to prevent the Jewish people from dying out altogether, a council of sages meeting at Lydda decreed that, to save his life a Jew might break any laws of the Torah except those which forbid idolatry, murder, adultery, and incest. This ruling became of vital importance in later persecutions.

The persecutions by Hadrian had, by death, and by emigration, especially to Babylon, greatly reduced the numbers of Jews in Palestine. Following his death in 138 his successor, Antoninus Pius, revoked many of the penal laws and sanctioned the setting up of a new Sanhedrin at Usha, and new schools were opened under Rabbi Simeon ben Gamaliel. His son and successor Judah I realized that peaceful conditions would not last indefinitely and that no single academy could hope to hold together for ever the scattered Jews of the world, and he set about the completion of a written instrument to take the place of oral tradition. This was to be, and indeed became, the definitive compilation of all traditional teachings in Mishnah form. This rapidly became accepted, both in Palestine and in Babylon as *the* Mishnah and became the authoritative work, second only to the Scriptures, for study in schools.

Life in Palestine did indeed, as Judah foresaw, become more and more difficult and, as the world centre of Judaism, Palestine gradually gave way to Babylon where, at Sura, Abba Arika, a disciple of Judah, set up a famous Mishnah school, and others were instituted at Susa and Pumbeditha. Further discussion and literary work at these schools produced a completion of the Mishnah and a commentary upon it, which together constitute the Babylonian Talmud; this was subsequently further edited and refined over a long period. The later history of the Babylonian Jews is considered in Chapter 5.

Religious and intellectual activity had not however ceased in Palestine and in the third century A.D. the Rabbi Jothanan ben Nappacha founded the academy at Tiberias on the Sea of Galilee, which produced the Palestine Talmud.

Following its production, conditions rapidly deteriorated in Palestine. The Byzantine Christian Church now held supreme secular power and repressed Judaism in every possible way. The Judaean Sanhedrin found it impossible to meet regularly to fix the dates of festivals and in 425 the Patriarchate was abolished altogether by the emperor. Shortly before this the Patriarch Hillel, recognizing that the Sanhedrin could no longer exert any effective authority, prepared a permanent calendar of fasts and festivals, based on astronomical calculations, for the use of world Jewry. This unselfish act proved to be of great benefit not only immediately but for many subsequent generations.

Literary activity however continued in Palestine for many centuries and indeed never ceased completely, though to outward appearances the country became henceforth for more than a thousand years the Holy Land solely of Christians and Muslims, and in Jerusalem Jews were admitted only to the Wailing Wall, the outer wall of their temple precinct.

Literary work after the abolition of the Sanhedrin continued to be based on Tiberias, and the greatest achievement of subsequent centuries was the establishment by the Massoretes of Palestine of the Massoretic text of the Scripture, which is still the text used by Jews throughout the world.

In the ninth century a Gaonate was established in Palestine which promoted the study of Palestinian literature, and maintained relations with Jews not only in adjacent countries but also in Italy where schools of Judaism were set up, the first to be established in a Christian country. It appears that in the Dark Ages, whereas Babylon provided the channel of communication with the Muslim world, Palestine, despite all obstacles, was the land from which Christian Europe maintained contact with eastern Judaism. This fact may be of some importance also in tracing the physical movements of the Jews themselves in Europe during this obscure period. Epstein (1959) quotes traditions that Charlemagne actively promoted Jewish scholarship based on Palestine in his dominions.

Tiberias remained a Jewish centre throughout the Middle Ages, including the agony of the Crusades, and indeed has survived as such until the present day.

THE FORERUNNERS OF ZIONISM

There has always been a mystical element in Judaism, but this underwent a special development in Europe in the Middle Ages around a movement known as the Kabbalah, which found particular expression in a work known as the Zohar, a mystical commentary on the Pentateuch perhaps dating from as early as the second century, but only becoming well known and influential in the thirteenth. It is mentioned here because there arose in the years 1550–90, at Safed in the hills of northern Galilee, a number of centres of Zohar teaching which may have had a considerable effect on the subsequent history of the Jews; not only did they provide a centre from which mystical teachings and literature spread back to Europe, but they established a fixed point in Palestine for another religious development, that of Hassidism, which, originating in the Ukraine in the eighteenth century, spread widely in Europe and led Mendel of Vitebsk to take a party of three hundred Hassidim to settle in Safed and Tiberias.

All these movements were, in fact, developments of a tradition which had existed throughout the Middle Ages, of elderly Jews migrating to Palestine to live, die, and be buried in the land of their fathers. Modern Zionism has undoubtedly been influenced by this tradition, but the immediate cause of the modern movement, and the main drive behind it in its formative years, was the pogroms which took place in lands under Russian control following the assassination of the Tsar Alexander II in March 1881.

In this book as a whole we are examining the Jews of all the world and, from a purely scientific point of view, it is most convenient that we have in Israel samples for study from nearly all the Jewish communities of the world. We also now find in Israel many who were born there of recent migrants, but, as already stated, their study is a task for the future.

In this chapter we are interested primarily in the genetical composition, if it can be ascertained, of the original Jewish inhabitants of Palestine. As we have seen, there was never a time when there were no Jews in Palestine, but at some periods there were very few, and it is questionable whether it would now be possible to assemble enough of their unmixed descendants to allow a meaningful genetical survey to be made—

except in the case of the Samaritans. It is, however, worth while to look back to the time, perhaps about A.D. 400, when movements of Jews out of Palestine had almost ceased, and then to look back from this standpoint still further and to consider what may have been the genetical composition of the successive bodies of migrants that, over the previous thousand years and more, had moved out to almost every part of the known world.

We see that the Diaspora can be divided into a number of stages. First came the enforced deportation of captives from from the north to Assyria and beyond, partly balanced by the sending of foreigners to replace them in Samaria. Next came the deportations from Jerusalem and Judaea to Babylon, and the flight of many elsewhere, especially to Egypt. Only a small proportion of the Babylon captives returned at the time of the rebuilding of Jerusalem and the Temple. Over the next thousand years Babylon became more and more recognized as a second home to Jews, and ultimately replaced Jerusalem and Palestine as the cultural and religious centre of Judaism. During the periods of Greek and Roman dominance, departures from Palestine were perhaps especially towards Egypt, but probably also to Babylonia, and certainly there were innumerable movements both direct from Palestine, and from the primary centres of dispersal, to all the lands bordering the Mediterranean.

Apart from a few sets of skeletons we have no means of knowing the genetical composition of the various groups of Jews of Palestine at the time of their successive departures. Even in such a small country there probably were slight differences between the north and the south. More important, however, is likely to have been a gradual incorporation of genes from outside the country, so that the earlier emigrants were probably nearer in their gene frequencies to the original Israelites or Hebrews than were later emigrants. Of possible genes from outside, those from Negro Africa are the only ones easily recognizable. As will be shown in later chapters, these are to be found in many Jewish populations, and there are indications of increased frequency in later emigrants. Existing non-Jewish authochthonous populations of the Near East may give some indication of the composition of the Jewish population as it was before Palestine was almost completely depopulated of Jews.

4

THE SAMARITANS

THE parable of the Good Samaritan is one of the best known of Bible stories, and several other New Testament stories about the Samaritans are also familiar. Few people other than Jews in Israel, and Christian theologians, realize, however, that the descendants of the biblical Samaritans still exist as a community and that their priests still maintain, as did the woman of Samaria, in speaking to Jesus at Jacob's Well, that their holy Mount Gerizim has priority over Jerusalem as the centre of Jahweh worship. Indeed in 1963 I had the experience of visiting Jacob's Well and the summit of Gerizim with the Samaritan Chief Priest, and being given booklets written by him supporting this very doctrine. The well is now in the crypt of an incomplete modern Christian church, but in this arid country wells are much more permanent topographic entities than buildings, and it is not unlikely that this same well does go back to Jesus and to Jacob.

The Samaritans are a distinct Jewish sect, now numbering some hundreds of individuals, living in Nablus, and in Holon near Jaffa, both now incorporated in Israel. They accept the Torah or five Books of Moses, and the earlier chapters of the book of Joshua, but reject the rest of the orthodox Jewish Bible (the Old Testament) together with such subsequent literature as the Talmud and the Mishnah.

They possess detailed and continuous records of their own community which go back further than those of almost any other culturally and genetically well-defined community in the world.

They indeed claim such continuity going back to the time of the Patriarchs and believe that their oldest copy of the Torah, the Abisha scroll, was written in patriarchal times.

THE ORIGIN OF THE SAMARITANS

Until a few years ago both the Samaritan leaders and most Jewish authorities agreed in identifying the northern Israelites of the time of the deportations to Assyria as the direct forerunners and ancestors of the modern Samaritans. As members of the tribes of Joseph (the half-tribes of Ephraim and Manasseh) and of Levi they distinguished themselves from the 'lost' eight tribes who lived farther north, and they claimed that their religion of Jahweh and the Torah had never been contaminated by pagan syncretic additions by foreign colonists sent in by the occupying power to replace the deported natives. Jewish authorities, following the narrative of the book of Nehemiah, claimed that such contamination had taken place. One Jewish authority, Gaster (1925), however, while accepting the northerners of the time of Nehemiah as the ancestors of the modern Samaritans, accepted also the claims of the latter as to the purity of their religion, stressing the total absence from their scriptures and practice of any pagan syncretic element.

The discovery of the Dead Sea Scrolls has led to a thorough reassessment of the history of religion in Palestine during the last pre-Christian centuries, and Coggins (1975), in particular,

has reconsidered the position of the Samaritans. In the following account I have made considerable use of his work, but I am, of course, solely responsible for the conclusions here put forward.

The word 'Samaritans' occurs only once in the English Authorised Version of the Old Testament, and the accuracy of the transliteration is questioned by Coggins. The city of Samaria is of course a very ancient one, but it is not clear how far the name is connected with that of the Samaritan religious community, who are much more closely associated with the city of Shechem, about ten kilometres farther south-east. The name which the Samaritans apply to themselves is 'Samerim' meaning 'keepers' (i.e. of the Torah) from a root which is familiar in the modern 'Tel Hashomer'. The linguistic relations between a large number of similar names are critically discussed by Coggins.

The Old Testament, of course, contains numerous accounts of conflicts between Israel and Judah, and between the religious establishment in Jerusalem and the supporters of worship at other shrines. Though the Samaritans later claimed for themselves a continuity with some of these opponents of Jerusalem, such continuity is questioned by Coggins. The Apocrypha, which bridges the gap between the Old and New Testaments, contains, however, frequent mentions of Samaritans, and it is clear that these were the direct forerunners of the New Testament (and modern) Samaritans.

The discovery of the extensive library of the Qumran sect has thrown considerable light on the history of the various recensions of the Old Testament. Previously three main versions had been known, the orthodox Jewish Massoretic text, the Samaritan version of the Torah, and the Septuagint, a Greek translation prepared during the third century B.C. by and for the Greek-speaking Jews of Egypt, and subsequently adopted by the mainly Greek-speaking early Christian church.

The so-called Massoretic text now used by orthodox Jews was finally edited by the Massoretes of the seventh century A.D., but it is essentially that approved by the Synod of Jamnia, near Jaffa, about A.D. 100. Independent evidence that the present accepted text had been almost completely established at this time is given by the recent finding of manuscripts of several Old Testament books, certainly written before A.D. 135. These were discovered in 1951 in a group of caves at Murabba'at, some miles south of Qumran, together with documents associated with the Second Jewish Revolt against the Romans in A.D. 132–5, and a wealth of other documents including the oldest Hebrew document known—a palimpsest probably of the eighth century B.C.—and a letter signed by the historical leader of the Second Revolt, Simon ben Kosebah.

There was formerly a tendency, especially among Jewish scholars, to regard departures of the Septuagint from the Massoretic text as idiosyncrasies of translation. Those of the Samaritan Torah were regarded as very late and secondary, despite the fact that the manuscript is written in an archaic cursive Hebrew script, a style predating the square alphabet thought to have been introduced at the time of the editing of the Torah about 500 B.C.

However, all manuscripts known until recently were separated by hundreds of years from the original compilation of the versions concerned and none dated from before the Christian era.

The situation was completely altered by the discovery of the Qumran manuscripts which date mostly from the last two centuries before the Christian era. They show that there was then already a standard text similar to that of the Massora, but with considerable minor variation among the manuscripts, including some which give the support of Hebrew originals to departures of the Septuagint from the Massoretic text. There is also support, in the case of some manuscripts of the Torah, for most of the 'idiosyncrasies' of the Samaritan version. Taking into account the many Qumran versions, the Massoretic text, the Septuagint, and the Samaritan Torah, almost every possible combination of variant readings can be found, thus showing that while, broadly speaking, a standard text of the Old Testament was established by about 100 B.C., if not earlier, almost all the variant readings found in versions now existing were in circulation by the same date, and cannot be late mistranscriptions or, in the case of the Septuagint, mistranslations.

The way is thus open to an admission of the antiquity and authority of the Samaritan Torah, and hence of the antiquity of the Samaritans themselves as a continuing religious community, but the precise date of their origin as such remains to be ascertained.

Our main sources of information on the history of the Jews during the last five centuries before the Christian era. besides the Apocrypha, are the *Antiquities* of Josephus, and the *Chronicle II* of the Samaritans themselves, the 'Book of days, containing the events from the entry of Joshua the son of Nun into the land of Canaan up to the present day' (i.e. the seventeenth century A.D.). The section dealing with the period which we are now considering has only recently been published, and most of the history of more recent periods, which would have been of particular interest in the context of the present book, unfortunately remains unpublished. I have relied almost exclusively on the work of Coggins for a critical interpretation of these records.

The chronicle up to the time of Ezra and Nehemiah, though incorporating important independent traditions, is essentially a polemically selected and edited version of the orthodox Old Testament. This alone shows it to be a late compilation. It places the break with orthodox Jewry at the time of Eli and Samuel, but Coggins concludes that the Samaritans of the New Testament were but one of a number of sects which in the previous two or three centuries had arisen in antagonism to the Jerusalem Temple cult. Far from promoting a syncretic cult they, like the Qumran sect with which they had much in common, were concerned with promoting a pure and ascetic worship of Jahweh with a properly-constituted priesthood. Early Christianity was in a sense another such cult.

The Samaritans did however show one feature which distinguished them sharply from the Jerusalem Jews—their attachment to Mount Gerizim as their centre of worship—and perhaps the most decisive act of separation was the building of a Temple on the mountain in the third century B.C., traditionally with the encouragement of Alexander the Great, but this and the date are open to considerable doubt.

A further cave-find has particular relevance to the Samaritans. Early in 1963 I happened to fly from Jerusalem to Beirut in the company of Mr. Lankester Harding, the eminent archaeologist and Director of Antiquities for Palestine, and he told me of a new discovery of documents in a cave in the hills north of Jericho. The documents found at Wadi Daliyeh were papyri, mostly economic documents from Samaria, together with pottery and more than three-hundred skeletons. Coggins (1975) quotes G. E. Wright's suggestion that these were Samaritans who had fled Alexander the Great's punitive expedition against Samaria in 331 B.C. The enemy troops—on this view those of Alexander—evidently found the hiding place of the fugitives and smothered them by building a large fire at the mouth of the cave.

If however these were Samaritans in the strict sense (and not just people of Samaria) this would argue strongly against identifying Alexander as the ruler who encouraged the building of the Temple.

This Temple, whenever it was built, was destroyed in the second century B.C., probably by John Hyrcanus in 128 B.C. It is claimed that 160 years later, Pontius Pilate, who had already condemned Jesus to death, ordered his troops to massacre the participants in a purely religious march to Mount Gerizim, where a prophet had promised to reveal the ancient vessels of the Temple hidden in a cave. The protests at this crime led to his recall to Rome. This was also a time of extreme bitterness, and even murderous strife, between Jews and Samaritans. But finally, upon the suppression in A.D. 70 of the great Jewish Revolt, the Samaritans again suffered with the Jews, and it was almost certainly then that the ancient city of Shechem, near the modern Arab village of Balatah, was destroyed, to be replaced about 2 kilometres to the west by the new city, Flavia Neapolis, which has become modern Nablus.

A large part of the library of the Samaritans is said to have been destroyed in the time of Hadrian in A.D. 130.

POST-BIBLICAL TIMES

For most of the remainder of the Roman period Samaria had relative peace, and finally in the fourth century Baba Rabba actually set up an independent state. At this time as many as possible of the ancient documents were collected together and an attempt was made to preserve them. Since then there has been no serious destruction of records. At this time, too, there was a fresh flowering of poetic and religious writing, most of which has survived. A new temple on Mount Gerizin appears to have been built soon after the time of Hadrian.

Soon, however, came Byzantine Christian rule which for 250 years cruelly oppressed the Jews and also the Samaritans, so that there were frequent revolts of the latter, giving excuses for further oppression. In A.D. 474 the Emperor Zeno expelled them from Mount Gerizim, presumably destroying the Temple, and built a Church on the mountain in honour of the Virgin Mary, extensive ruins of which can still be seen. Perhaps the most serious revolt of the Samaritans was that of A.D. 529 under Justinian, when many Christians in various parts of Palestine were killed; it was probably on this occasion that the Church of the Nativity at Bethlehem was burnt down.

It is known that following the early Muslim invasion of Egypt numbers of the Samaritan community fled to an unidentified island in the Red Sea, known as Gezirat al-Samir (Samaritans' Island). It is possible that this is to be identified as Yotba (modern Tiran) at the mouth of the Gulf of Eilat or Aqaba, where about A.D. 500 (i.e. before the emergence of Islam) there was an autonomous Jewish community.

It was in the unhappy period of Byzantine rule that there began the gradual decline in numbers and influence of the Samaritans. Despite, or perhaps even because of, the relative moderation of the rulers during the succeeding Moslem period, this decline continued until about 100 years ago, but the Samaritans remained faithful, as they still are, to Jahweh, to the Torah, and to their sanctuary.

While the history of the Samaritans as a nation must perhaps be regarded as ending with Baba Rabba, it is mainly during the subsequent period that we get outside glimpses at the community from sources other than Jewish, which are of considerable interest when one attempts to reconstruct their demographic record. Moreover it is the sad story of greatly diminished numbers in the past two centuries that is especially important in the interpretation of modern biological studies of the community.

Apart from the question of changing population numbers, which must be considered in detail, only a few major events in the story of the Samaritans since the advent of Islam can be mentioned here.

In A.D. 1624 the priestly house claiming descent in the male line from Aaron died out, and the high priests have henceforth been drawn from men tracing descent from Uzziel, son of Kohath, son of Levi. The millenial conflict with the Jews, in the political sphere at least, may be regarded as having ended in 1841. In that year, when the Muslims threatened to exterminate completely the last remnants of the Samaritan people, they were saved by the chief rabbi of Jerusalem who gave them a certificate stating that 'the Samaritan people is a branch of the children is Israel, who acknowledge the truth of Torah'. This proved that they were 'people of the Book' and so entitled, like Jews and Christians, to protection. In 1854 they came under the protection of the British Consul. In 1948, when the State of Israel was founded, Nablus remained in Jordan, but the Samaritans shared with all other Israelites the right of immigration into Israel. This meant that girls from Nablus could marry men living in Israel, but Jordanian law prevented girls from Israel going to marry men in Nablus, where the whole of the priestly family remained. Samaritans from Israel were allowed to take part in the Passover celebrations on Mount Gerizim. Since 1967 there has of course been no legal bar to movement between Nablus and the other community at Holon near Jaffa, but the priestly family remains at the former place.

From a variety of sources it has been possible to find estimates of the numbers of the Samaritan community at various times. In the seventh century A.D. there are variously said to have been 80 000 or 30 000 Samaritans living at Caesarea. This was probably their main centre in Palestine outside Nablus; the total number of Samaritans at this time is therefore likely to have been of the order of 100 000.

In 1163 numbers (apparently of men or families) are said to have been 1000 in Nablus or Shechem, 200 in Caesarea, 300 in New Ashkelon in south-east Palestine, and 400 in Damascus. By the end of the thirteenth century there seem to have been about 1000 Samaritans in Nablus and perhaps another 1000 in neighbouring villages.

In 1488 the numbers are given as totalling 2500, including 500 in Egypt, and others in Shechem, Gaza and other Palestinian localities, and in Damascus.

In 1548 the Ottoman registers show only 52 families, and numbers continued to fall so that in the nineteenth century,

between 1838 and 1881, we find a series of low estimates fluctuating between the lowest, 122 in 1853, and 160.

In the seventeenth century there had been colonies at Jaffa and Gaza but these had almost, if not quite, vanished by 1800. From this period until 1948 the Nablus community was practically the only one but there may always have been a few families in Jaffa, and since 1948 there has been a small but flourishing community at Holon. Since 1900 there has been a steady increase in numbers, at first in Nablus and latterly mainly in Holon.

Even more important genetically than the low absolute numbers is the very low total of females, 51, reached in 1901 when there were 97 males. Since then the proportions of the sexes have evened out. Since 1900 the total population has more than doubled, to 343 in 1960, 209 in Nablus and 134 in Holon. There is still a slight excess of males, and there are said to have been three marriages of Samaritan men with Jewish girls in the Holon community.

Dr. Batsheva Bonné in 1966, using all available data on population numbers, and on consanguineous matings, calculated that the genetically effective breeding population size was 92 at the time of writing, and that in 1875 it had fallen as low as 39. If the figure of a total population of only 122 in 1853 is accepted, the effective breeding population would then have been only about 35.

It is thus highly probable that in the past two centuries there has been considerable genetic drift, and frequencies of genes in general may have diverged considerably from the probably much more stable values existing up to about 1500 when the total population was numbered in thousands.

Neither in human populations generally, nor in particular cases, do we know, however, what influence may be at work in the form of natural selection, and especially selection in relation to the total genome, to stabilize the fluctuations predicted on the simple drift hypothesis. Thus we cannot, simply because numbers have fallen low, regard present gene frequencies as unrelated to those of many generations ago, or the present frequencies as irrelevant to a study of the origins of the Samaritan people in the first millenium B.C.

ANTHROPOMETRY

Among the genetical characteristics of the population we must, of course, reckon not only the blood groups and other simply inherited chracters, but also the anthropometric characters. There are two main sources of information on the physical characteristics of the Samaritans, resulting from the work of H. M. Huxley in 1900 and 1901 (Seltzer 1940), and of Genna in 1933. We have, moreover, photographs going back as far as about 1880, which show that from then until the present day a highly characteristic physiognomy has persisted. It must, however, be admitted that all photographers, including myself, have tended to depict mainly the priests with their picturesque robes and full beards, whose family we now know to differ genetically from the non-priestly families.

As regards anthropometric characters all that can be done here is to summarize very briefly the findings of Huxley and of Genna, and to recommend those who may be sufficiently interested to consult the original works.

Genna (1938) examined and measured far more persons than Huxley, and he also classified his data by family names. In addition he includes data from previous authors, including not only Huxley, but also Weissenberg, Szpidbaum, and

Ariens Kappers. In addition to his massive body of tables of physical measurements Genna gives demographic data and photographs of heads in three positions.

The whole corpus of data, both these of Genna and the others, needs to be brought together and treated by modern anthropometric and computational methods. It would also be of great interest to see a comparison made between living Samaritans and the 300 skeletons from Wadi Dalayieh (p. 19). Since however Huxley's results, thanks to his editor, Seltzer, are much more fully summarized than those of Genna, they have served as the main basis for the section which follows. According to Huxley, skin colour is unusually light for a Near Eastern population. Genna finds, among non-grey adults, 84 per cent of chestnut brown hair, and 16 per cent with lighter colours, including 4 per cent red. Huxley gives a higher proportion of light colours. Eye colours are heterogenous, with a few examples of blue. According to Genna no examples of light hair or eye colour occur in the priestly or Kahin family; and he attributes the lighter colours found in some members of the other families as due to miscegenation. Hair is usually straight or wavy; facial hair in males in usually abundant.

The Samaritans are rather tall, with a mean height of 172 cm. The mean cephalic index is 77·91, on the borders of meso- and dolicho-cephaly. The cranial vault is remarkably high, with a mean height of 140·22 mm. Forehead and face are generally long, so that the Samaritans are perhaps the most leptoprosopic group in the Near East. The nose is distinctly long and moderately broad. The lips are rather thick. The ears are large and often project markedly.

The following final conclusions as set out by Seltzer are of interest though some of the terms used are now obsolete:

There is little doubt from the examination of the metric data as well as the photographs that the basic element in the Samaritans is that of the Iranian Plateau stock. This is evident in part by the extreme nasality, the extraordinary height of the cranial vault, the sugar-loafed contour of the head and by the great length of the face. From the above eye colour sortings it seems to be at any event suggestive that this element entered the group along with a Nordic strain accentuating its leptorrhiny and its leptoprosopy. The second racial element of importance is the so-called Atlanto-Mediterranean. The dark brown-eyed class perhaps shows this element in its greatest strength even here it is intimately associated with Iranian Plateau characteristics. The third element, and of lesser significance, is the Alpine strain. This seems to be clearly linked with the brown-eyed class in its brachycephaly, its greater lateral dimensions of head and face, in its tendency towards mesoprosopy and in its weaker leptorrhiny. Although this Alpine strain is present to some extent in the group which we earlier designated photographically as the 'Ultra-Samaritans', it is undoubtedly most strongly represented in the small residual 'non-Samaritan' group.

. . . the Samaritans differ very markedly from the Maronites, Nusairiyeh and Druse series. These groups are hyperbrachy-cephalic in contrast to the Samaritans who present a low meso-cephalic cranial vault. This brachycephaly is attributed in the Maronites and Druse to the very strong elements of Alpine as well as Armenoid-Iranian Plateau blood. The Alpine strain is particularly of very minor significance among the Samaritans. There is also no close relationship between the Samaritans and the Bedawins, Akeydat, Maualy, Moslems and Turkomans. In the latter groups the dominant elements in varying strength is the Mediterranean-Arab strain, while in the Samaritans the significant stock is the Iranian Plateau. The Samaritans do possess strong Mediterranean blood, but whereas in the aforementioned groups it is Arab-Mediterranean, in the Samaritans it is the Atlanto-Mediterranean.

BLOOD GROUPS

The small numbers of Samaritans tested is, of course, a reflection of the smallness of the total population. Moreover, the Samaritans themselves maintain their tribal affiliations, some families claiming to belong to the half-tribe of Ephraim and some to that of Manasseh. The priestly family of Kahin, belonging to the tribe of Levi, maintains a partial endogamy within the endogamous total population. Therefore, despite the extremely small sub-total tribal numbers, the records of the different tribes have been kept separate in our tables.

Specimens collected at Nablus before 1958 cover the whole community (except in so far as deliberate selection may have been made). Between 1948 and 1967 the Nablus community included an enhanced proportion of the Kahin family, none of whom, however, moved to Holon. Hence the specimens tested by Dr. Bonné (now Mrs. Bonné-Tamir) in Israel in 1970 included no members of that family. Subsequently specimens were collected by her (Bonné-Tamir 1976) at a Passover celebration of the whole community on Mount Gerizim. These include members of all tribes and families. They are later to be analysed in detail but Mrs. Bonné-Tamir has kindly allowed us to quote the unpublished complete (and therefore mixed) data.

It has been possible to show by comparing the results of Younovitch in 1938 with those of Ikin et al. in 1963 that over a period of one generation the Kahin family has maintained a much higher A frequency than the rest. The overall gene frequencies obtained by adding all results together and ignoring the unknown degree of overlap are: A, 17%; B, 7%, but this conceals a distinct difference between the Kahin family with: A, 31·5%; B, 5·6%, and the rest with A, 13·1%; B, 7·8%. The A frequency in the Kahin family is unusually high for the region, but similar to that found in the Armenians. Frequencies of the A_2 gene are consistently remarkably high. Because of the differences in A frequencies, which shows that families are genetically different, the few data classified by families have been calculated separately also in the case of other systems.

For the MN system the M gene frequency is consistently distinctly low for the region, at 43 per cent. For the S gene frequency, not surprisingly with such low total numbers, calculated frequencies differ, but the average is about 35 per cent. The African Henshaw marker gene appears to be absent. The P_1 gene frequency is rather low, about 22 per cent. Rh-negative phenotype frequencies are consistently rather high for the region, at 15·7 per cent (d gene, 39·6 per cent). Tests for the other Rh antigens show little that is noteworthy. The cDe complex (which is common in Africans) is absent, or present only in low frequency, and the African V marker gene was not found. The main difference between the tribal groups is in the frequency of the D^u gene, and this might be an artefact.

The Kell gene K is absent, as is the African marker gene Js^a. The frequency of the Fy^a gene is also low. The African marker Fy^4 appears to be absent, but could be present in heterozygous form. The Mongoloid marker gene Di^a was not found. There is disagreement between series as to the frequency of the Lu^a gene and further tests are desirable.

The frequency of the haptoglobin Hp^1 gene, 40 per cent, is somewhat high for the region; that of Gc^2, 20 per cent, is rather low. Glucose-6-phosphate dehydrogenase deficiency, present in most other Jewish populations, has not been found. No data are available on the frequencies of other red-cell enzyme variants. The frequency of the phenylthiocarbamide taster gene, 75 per cent, is unusually high for any Old World population. The frequency of colour blindness in males, 27 per cent, at Holon, is exceptionally high, but only 101 persons were tested.

The incidence of congential abnormalities, especially neuromuscular ones and deafness, at Nablus appeared to me unduly high but I was unable to complete arrangements for expert diagnosis or treatment. Both of these will no doubt be made available through the national health services of Israel.

The genetic situation may be summed up by saying that the broad picture conforms to regional gene frequencies, but with some marked deviations which are almost certainly due to genetic drift in this small inbred community. African marker genes, present at low frequencies in almost all Near Eastern populations, are consistently absent. Thus, though even during the present century a very few non-Samaritan wives have been introduced into the community, the claim of the Samaritans to have avoided intermarriage with out-siders appears to be very largely true. The Samaritans can thus justifiably claim that they are as representative a sample as any population now living of the Israelites of biblical times.

5

THE ORIENTAL JEWS

THE term 'Oriental Jews' is commonly used to include all Jews in or coming from countries in Asia. In this chapter, however, we do not include the Yemenite Jews, who are described in a separate chapter and whose genetic composition differs greatly from that of all the rest of the Jews of Asia, themselves quite heterogeneous as far as their blood groups are concerned. Among this large family we shall first consider two branches whose history appears to go back to Old Testatement times.

THE LOST TRIBES OF ISRAEL: THE KURDISH JEWS

In 732 B.C. Tiglath Pileser, king of Assyria, captured Damascus and deported many of the inhabitants, who included members of the Israelite tribes of Reuben and Gad and the half-tribe of Manasseh, to the cities of the Medes. These tribes had been allotted, traditionally by Moses himself, lands to the east of the Jordan, including the land of Gilead, though, according to the bibilical narrative, they had to take their part under Joshua in the conquest of the lands on the other side of the river on behalf of their brethren of the other tribes. In 722 Shalmaneser V, the next Assyrian king, captured the city of Samaria and his successor Sargon II took away many inhabitants of the west bank to 'Halath, Hathor and the cities of the Medes'. This corresponds roughly to modern Kurdistan. Apart from some priests who were later sent back to continue the worship of Jahweh in Samaria, the Bible tells us little about the subsequent history of these groups of Jews. Isaiah refers to them as 'lost in the land of Assyria', and in Jewish tradition they are indeed generally regarded as 'lost'. Ben-Zvi (1958) quotes the apocryphal Book of Tobias as mentioning families of the tribe of Naphtali living among the cities of the Medes.

For a great many years after this we hear almost nothing of these lost Israelites, but it is perhaps to them, or perhaps to Babylonian influence, that we must attribute the conversion to Judaism of Queen Helen and the rest of the ruling house of Adiabene, a little kingdom within the Parthian Empire, in northern Mesopotamia, in the very early years of the Christian era. The royal mausoleum in Jerusalem is well known.

The next reference cited by Ben-Zvi is an account written in the second half of the twelfth century A.D., by the traveller Benjamin of Tudela who states that:

> . . . in the hills of Nisbur there are four tribes of Israel, namely the tribes of Zebulun, Dan, Asher and Naphtali, all descendants of the first exiles who were carried to this country by Shalmaneser king of Assyria.

Shortly before this the Kurdish Jews were involved in the unsuccessful rebellion of Jews in Persia, led by David Alroy, against their Persian overlords. The modern Kurdish Jews rightly claim him as their own hero, but it has been suggested that he was by origin a Khazar (q.v.).

The Modern Kurdish Jews speak Aramaic, which suggests that they have throughout the ages maintained some contact with the Babylonian Jews, who speak a similar dialect.

They know the Bible, but have very little knowledge of the Mishnah and the Talmud, despite the fact that these are largely the product of their neighbours the Babylonian Jews, and are indeed written in Aramaic.

In recent times the Kurdish Jews were largely tillers of the soil, as serfs of their Kurdish overlords. They were distinguished from the latter by their synagogue worship and the main Jewish rules regarding food and purity, as well as by their language, but their social life showed strong Muslim influences especially in the extreme subservience and seclusion of women.

The Kurdish Jews regard themselves as descended from the exiles from Samaria, and there seems to be no good reason to doubt this. As will be shown later, they are genetically distinct from the Babylonian Jews. Perhaps more open to doubt are their claims to locate in Kurdistan the tombs of the prophets Azariah, Daniel, Hananiah, Jonah, Mishael, and Nahum.

They have however suffered not only from being Jews among a Muslim population, but also from the disabilities of the Kurds themselves. The latter speak dialects of a single language, akin to Persian, and regard themselves as one nation, but they have never achieved independence. At present they are split between Turkey, Iraq, and Iran. Turkey does not allow the term Kurds but calls them 'Mountain Turks'. In Iraq they form a large compact population in the north and have repeatedly but with little success tried to achieve autonomy. In Iran, where they are linguistically and racially akin to the Persians, they suffer relatively little in the way of ethnic discrimination.

There were Jews living among the Kurds in all three countries, most of whom have now moved to Israel.

Not only the Kurdish Jews, but many other Hebrew communities in Asia, and some in Africa, have claimed descent from the 'lost tribes' and there is, indeed, much evidence to support many of these claims. It has, however, been suggested that these assertions, and the tendency of those making them to call themselves Israelites rather than Jews, may be due in part to a wish to avoid difficulties with their Muslim neighbours, for the latter have sometimes in the past tended to react differently to the two terms.

THE JEWS OF BABYLON

In Chapter 3 we have followed the history of the Jews of Babylon, a community founded at the time of the Babylonian capitivity in 586 B.C. For many hundreds of years there was the closest contact between them and the parent community in Palestine, so that it is unnecessary to repeat here the account of their joint history up to the abolition of the Palestinian Exilarchate in A.D. 425.

It is likely that at this time the Babylonian Jews, were, genetically speaking, still essentially displaced Palestinians, though probable that, in nearly 1000 years of exile, they had to some extent intermarried with the people of the land. Most of these autochthones would have been akin, linguistically and probably to a large extent genetically, to the Palestinian exiles, but probably in the course of time they incorporated substantial numbers of the Persian ruling race, and thus some genes of Persian origin are likely to have been incorporated also in the Jewish community.

For a short time after the suppression of the Palestinian Exilarchate the Babylonian Jews continued to enjoy religious freedom, but in the latter half of the fifth century they suffered growing persecution by the representatives of the Persian fire-worshippers who were now in power, and the Academies of Sura and Pumbeditha were closed.

Soon after the arrival of the Muslims and the foundation of the City of Baghdad by the Khalif Al-Mansur (A.D. 754–75) the Jews were once more in favour. They were regarded by Muslims as akin to themselves in race, language, and religion, but they had a long tradition of learning which the Muslims lacked and which they could supply. The heads of the reopened Academies thus achieved a widely recognized authority not only in Baghdad but throughout a large part of the Muslim world. Since a large part of the Mediterranean was now in Muslim hands students came from far and wide, both from Muslim and from Christian lands to the Academies of Sura, Pumbeditha, and Nehardea.

There was also a new flowering of biblical scholarship both in Babylonia and in Palestine, leading to the final redaction of the Hebrew Bible by the Massoretes of both lands, and to the production of alternative Massoretic texts, that of the Palestinians at Tiberias being the one finally accepted by the whole of Judaism. In about A.D. 930 an Egyptian Jew became Gaon (head) of the Academy of Sura and wrote a work *Emunoth Wedeoth* (Faith and Knowledge) which is regarded as epoch-making in Jewish theology. With his death, world Jewish leadership began to pass from Babylonia to Spain where it remained until the expulsion of the Jews in 1492.

THE PERSIAN JEWS

There have almost certainly been Jews in Persia continuously since very soon after the first deportations to Assyria, but it is only at rather long intervals that history gives us glimpses of them. Almost inevitably the Jews of Persia must have been recruited from the communities in Kurdistan or Babylon or, more likely, from both of these. However, since the modern Jews of Persia, unlike the Babylonian and Kurdish Jews, speak a Judaeo-Persian dialect and not Aramaic, these Persian Jews must have a long separate history of their own.

According to authorities quoted by Ben-Zvi (1958) there were in the twelfth century about 60 000 Jews in the country, while at the time of the foundation of the state of Israel there were rather more than 100 000 of them, concentrated in a few of the largest towns.

Ben-Zvi (1958) thinks that the earliest Jews of Persia were drawn from the captives taken by Tiglath-Pileser in 732 B.C. from among the Israelites on the east side of Jordan. One of his reasons is that the village of Gillard on the outskirts of Tehran, the site of an ancient Jewish cemetery, is said to take its name from the land of Gilead, part of the territory of the tribes of Gad and Manasseh east of Jordan. He cites many other traditions found among Persian Jews, notably those of Demavend, in support of this view. He further states that some Jews from Persia claim to belong to the tribe of Benjamin, implying descent from the persons deported from

Israel in A.D. 721, while others claim to belong to the tribe of Judah (involving descent from the Babylonian captives of 586 B.C.). He also says that many communities in Kurdistan and in the rest of Persia, now Muslims, have traditions and customs showing Israelite ancestry.

The story related in the biblical book of Esther cannot be dated precisely and may be largely legendary, but, if it contains a kernel of historical truth, it means that, at some time during the Achaemenid period (559–331 B.C.), there were in Persia numerous important colonies of Jews, perhaps extending far east into modern Afghanistan.

The Sassanide dynasty of Parthian rulers of Persia, who held power from A.D. 227–651 were fanatic supporters of the Zoroastrian religion of fire-worshippers and persecutors of unbelievers. Persecution was particularly fierce under Yazdegerd II and Firuz in the fifth century and this led to considerable emigration of Persian and Babylonian Jews to join the already established colonies in central Asia and Bukhara.

Other Sassanide kings showed favour to the Jews. Shapur (A.D. 310–79) brought several thousand Jews from Armenia to Persia, and some of the kings married Jewish queens. There are known to have been two temporarily successful Jewish revolts against the Sassanides.

Very early in the spread of the followers of Mohammed, Persia came under their sway, the ancient capital, Ctesiphon, being taken in A.D. 637. The Muslims proved, on the whole, to be more tolerant than the Zoroastrians, but there were from time to time, nevertheless, persecutions and revolts of the Jews. The revolt of Kurdish and Persian Jews under David Alroy in the late twelfth century A.D. has already been described.

One period of extensive assassination and persecution began in Meshed as recently as A.D. 1839. An incident involving the killing of a dog on a Muslim fast day was exaggerated and distorted, so that a Muslim mob broke into the Jewish quarter and killed 32 Jews. Following this the whole Jewish community, on pain of death, was forced to profess conversion to Islam, but, like the Marranos of Spain under the Christian regime, they continued to practise the Jewish religion in secret, while following that of Mohammed in public. Many fled to such places as Bukhara and Afghanistan, and even to India and England, but the majority remained in Persia until at least half of them, about a thousand persons, were able to migrate to Israel.

THE JEWS OF BUKHARA

The Jewish community in Bukhara and Turkestan use the same Judaeo-Persian dialect as those of Persia. They claim to be descended from the lost ten tribes and are undoubtedly a very ancient community. Many Jews almost certainly fled from Persia to Bukhara at the time of the Zorastrian persecutions in the fifth century A.D.

The Jews of this region appear to have survived the Muslim conquests and conversions without serious detriment, and to have shared in the general prosperity in the twelfth century, as described by Benjamin of Tudela. In the next century the region was devastated by the Mongol invaders. The latter soon became Muslims and then persecuted both pagans and Jews on religious grounds. The Bukhara community was re-established at the beginning of the fourteenth century, when more migrants came in from Babylonia and Persia, and another episode of prosperity and literary activity followed. When, in 1598, the Samarkand community was destroyed by the victory of Bab Mohammed Khan over Shah Abbas of Persia, many of the Jews who were there fled to Bukhara.

In the eighteenth century the community was much influenced doctrinally by Sephardic and Palestinian itinerant rabbis.

The annexation of Bukhara in 1865–6 by Tsarist Russia led, after initial restrictions, to a time of prosperity for the Bukhara Jews, and, with the opening of communications with the rest of the Tsar's dominions, the Jews of Bukhara in 1802 established correspondence with those of Lithuania. Unlike the Ashkenazim in European Russia, the Bukhara Jews were not subject under the Tsars to pogroms or severe judicial restrictions, but conditions for them deteriorated with the advent of Soviet rule.

Emigration to Palestine began in 1868 and continued until about 1936, by which time the number of Bukharans in Palestine had reached 2500. Much larger numbers still live in Bukhara.

JEWS BETWEEN THE CASPIAN AND THE BLACK SEAS

In the Soviet Union, north of the Iranian frontier and west of the Caspian Sea, are a number of groups of Jews, distinct from one another as well as from the Ashkenazim from farther north in the Soviet Union. Ben-Zvi regards all these as derived primarily from the deportations of 719 B.C. from northern Israel, with later additions from the already dispersed communities of Kurdistan, Babylon, and Persia, and from the Parthian-Jewish kingdom of Adiabene.

We are given little separate information about the Jews of Azerbaijan. They are presumed to be essentially Persian Jews separated by the accident of the position of the modern frontier.

The Jews of Armenia have a long history. We have already referred to the moving by King Shapur of Persia in the fourth century A.D. of 50 000 Jewish families from Armenia to Persia, implying a very large community in the former country in earlier times. Ben-Zvi quotes Flavus Josephus regarding the settlement of many Jews in Armenia in the reign of King Tigranes, a member of the Herodian family, about the year A.D. 10.

There are still Jews in Armenia, who continue, like the Kurdish Jews, to speak Aramaic. There is now virtually no contact between Jews remaining in Armenia and those outside the Soviet Union.

The Jews of Georgia form a distinct group, speaking Georgian for everyday purposes and writing in the Georgian left-to-right script and not in Hebrew characters. Their personal names, also, are essentially Georgian, but they maintain a knowledge of Hebrew.

Migration of Georgian Jews to Palestine began in the 1860s and there is now a substantial community of them there. In 1926 there were 21 000 Georgian Jews in Caucasia. In 1933 a museum was set up in Tiflis, devoted mainly to the history and ethnography of the Georgian Jews, and containing also Karaite records. Ben-Zvi states that a report by the Director, A. Krikhely, appeared in *Soviet Ethnography*, 1946, No. 4, p. 219. There is however now virtually no contact of the Georgian Jews with the outside world.

Some Christian communities in Georgia have traditions indicating that their ancestors were of the Jewish religion.

The Jews of Daghestan, including the Mountain Jews of the Caucasus, speak a Persian dialect known as Tatti. Shortly before the First World War they are said to have

numbered 25 000. They were scattered through numerous villages, living in very large households, and many of them were farmers. They are said to be of particularly impressive physique.

When, in the nineteenth century, Tsarist Russia annexed the area between the Caspian and Black Seas, the Jews of this area established contact with the Ashkenazim of Poland and Lithuania, and until 1947 they are known to have maintained their Jewish customs, but little is known of their present condition.

THE JEWS OF AFGHANISTAN

In 1948 there were rather less than 8000 Jews in Afghanistan, speaking the same Judaeo-Persian dialect as those of Persia. Ben-Zvi quotes Benjamin of Tudela as stating that there were in Afghanistan in the twelfth century A.D. tribes of Jews claiming descent from the northern Israelite tribes of Dan, Zebulun, Asher, and Napthali, Ben-Zvi himself, quoting various authorities, regards the whole large and important Pathan nation as of Jewish descent.

THE JEWS OF INDIA

The Jewish community of Malabar is recorded as being set up by Joseph Rabban in A.D. 379 and appears to have continued as a self-governing state until 1471. It was visited by Marco Polo just prior to 1300. It was conquered, apparently by a neighbouring Indian ruler, in 1471 and the few remaining Jews removed to Cochin where their descendants remained until their recent migration to Israel. They were divided into two endogamous groups, the larger one of 'Black Jews' and the smaller, socially dominant, one of 'White Jews'. The early history of the community is recorded on brass tablets on the walls of their synagogue of Cochin. They are said to have had written records of numerous other Jewish communities in Asia, including those of China (see below).

Another ancient Jewish community in India is that of the Bene-Israel Jews of Bombay. Various accounts have been given of their history. Like many other groups of Jews in Asia they claim to be descended from the ten lost tribes. One account of their arrival in India, quoted by Stritzower (1971) from Kehimkar, is that they were wrecked on the coast of the Konkan in western India not far from Bombay. Most of the travellers were lost together with all their possessions but seven couples founded the present community which maintains a few of the basic Jewish rites and customs.

Another tradition about their departure from Palestine is that they left following the profanation of the Temple by Antiochus Epiphanes in 175 B.C. They are said to have been restored to orthodox Judaism following the visit from an itinerant rabbi, David Rababi, in the twelfth century A.D.

In the early twentieth century there were 10 000 Bene-Israel Jews in Bombay and some 3000 elsewhere in India. Some 10 000 had emigrated to Israel. Those in Bombay resembled local Indians in personal appearance, and followed them in dress, in speaking Marathi, and in many social customs. Many also spoke English. They had, until recently, like the Cochin Jews, been divided into Gora or White Bene-Israel and Kala or Black Bene-Israel.

The Baghdadi Jews of Bombay came as merchants from Persia about A.D. 1836. They speak a Hebrew-Persian dialect and are much lighter in colour than the Bene-Israel with whom they very rarely intermarry.

THE JEWS OF CHINA

There are various references to Jews in China in ancient Chinese writings, and Marco Polo, at the end of the thirteenth century A.D., mentions Jews at the court of the Great Khan in Khan-baliq (Peking). However, the only ancient community of Jews in China about which we have any detailed information is that which, until the early years of the present century, existed at K'ai-fêng Fu in central China, the history of which has been written by White (1966). Though the last members of the community have become merged in the Chinese population, the records of its history from A.D. 1163 onwards are remarkably full, in the form of stone inscriptions, the written records of the community itself, and descriptions by Christian missionaries from A.D. 1605 onwards. It has also been possible to supplement the community's own genealogical records from official Chinese lists of civil and military appointments.

There were five historical inscriptions in the synagogue area, carved in Chinese characters on three stone tablets. On the back and front of one tablet were inscriptions with dates equivalent to A.D. 1489 and 1512 respectively. A second tablet had two inscriptions, one dated 1663 and the other undated and perhaps slightly later. A third had one inscription only, dated 1679.

The first and third of these stones were given by the Jews to the authorities of the Anglican cathedral in K'ai-fêng in 1912 and subsequently, in order to satisfy government officials, were formally sold to the Anglicans on condition that they should never leave the province. They were set up on either side of the western entrance to the cathedral and, as far as is known, they are still there. The present location of the other stone is not known but a rubbing of one face and a transcription of the other have been preserved.

The inscriptions on these stones are the main authorities for the history of the community. There is, however, considerable doubt both as to whence these Jews came and when they arrived. They are said to have come from 'T'ien-chu', which usually means 'India'. Some authorities take this literally and variously suggest that they came by sea from Cochin or overland from north-west India. White however thinks that the name, in this context, means Judaea or Syria. He stresses the fact that the Hebrew writings of the community contain Persian letters of the alphabet and many Persian words, which point to a long sojourn of the ancestral community in Persia. There are also passages in Aramaic which has remained the language of the Babylonian and Kurdish (though not the Persian) Jews. He also draws attention to the historical and archaeological evidence for the early presence of Jews at numerous points on the 'silk route' from the Near East through Persia to China.

Seventy-three clans with over five hundred families are said to have entered China during the Sung dynasty who reigned in K'ai-fêng Fu from A.D. 960–1126, and to have been invited by the Emperor to settle there.

White points to the use by the community of the calendar of the Seleucid era, or 'Era of Contracts' rather than the 'Era of the Creation', suggesting that they left Persia before the tenth century A.D. A synagogue ('Temple of Purity and Truth') was first built in A.D. 1163. It was rebuilt in 1279 but was destroyed by floods in 1461 and again rebuilt more elaborately than before. It was further extended at some time before 1488. It was presumably this rebuilt and extended synagogue which was first seen by Jesuit missionaries in the early seventeenth century.

In 1642 the city was besieged by rebel troops who breached the dykes of the Yellow River so that floods again destroyed the synagogue, as well as most of the scriptures that were kept in it.

A member of the Jewish community, Li Kuang-T'ien, became celebrated for his heroism in the defence of the city, and White gives the following translation of a contemporary eulogy written by Sun Su-Ch'ien (who appears not to have been a Jew).

> He of the flowing beard
> Rebirth of Sui-yang,
> Again saves the fateful town;
> Men's trusted chieftain
>
> The Ming reign extended,
> Through merit of valour.
> Who is this Gentleman?
> Only a scholar!

Some scriptures were recovered from the waters and the synagogue was rebuilt a few years later; we have detailed drawings of it, made by Père Jean Domingue, S.J., in 1722.

The cost of the rebuilding was probably a considerable strain on the resources of the community. This and the loss of their library may have started their downhill course. By 1851 they were poor and though the buildings were still standing they were dilapidated. There had also been much religious persecution and very few people could read Hebrew, so that morale was low, and soon, because of the urgent need for money to keep themselves alive, they began to sell the stones and other materials of the synagogue for use in the adjacent Muslim mosque, and in a few years the building had disappeared.

The scholars of the community took to studying Chinese rather than Hebrew literature, so that they could take the official examinations, leading to well-paid posts in the civil service. Today not only has the synagogue gone but the community itself has disappeared into the surrounding Chinese population.

However, most of the written records which were still in existence in the nineteenth century, and some other objects, have been preserved in the museums of America and Europe. Apart from the inscriptions, in Chinese characters, already mentioned, nearly all of the documents are in Hebrew. One is in Hebrew and Chinese with some passages in Aramaic.

The present, or last known, locations of most of these documents and objects are recorded by White, but the statements are somewhat scattered, mainly in Part II of his book. A single cross-referenced index of documents and of locations would be most useful.

The two largest collections appear to be those of the Royal Ontario Museum, Toronto, Canada and the Hebrew Union College, Cincinnati, Ohio, U.S.A. In England some documents are in the David S. Sassoon Collection in London, the British Museum Library, the London Jews Society, 16 Lincoln's Inn Fields, W.C.2, the Bodleian Library, Oxford, and the University Library, Cambridge.

There is now no possibility of carrying out genetic studies on living descendants of members of the community, but photographs of many of them exist and are reproduced by White; it would be interesting for these to be examined by a physical anthropologist. To a layman most of them look like normal Chinese subjects, but this impression may be due in

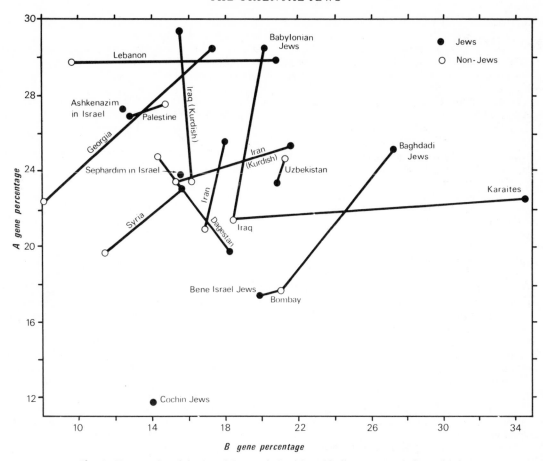

FIG. 2. Frequencies of the *A* and *B* genes in Jewish and indigenous populations of Asia.

part to their Chinese costumes. Some outsiders who had met members of the community have stated that they differed visibly from the Chinese.

One important piece of evidence of ancestry is the multi-lingual document already mentioned, which was written at about the time of the flood of 1642. The greater part of it is a genealogical record, and it shows that many women with non-Israelite surnames married into the community. It is probable, also, that many non-Israelite girls were adopted by families of the community. Marriage of women of the community to outsiders was, however, forbidden.

It is of interest that, in the early seventeenth century, there existed in K'ai-fêng, besides the Israelite community, an old Muslim one to which reference has already been made, and which still exists. There was also an old Christian community, then in the last stages of distintegration.

It is idle to speculate on what might have happened had the Israelite community maintained its integrity, but almost certainly, if it had lasted another fifty years, some of its members would have migrated to Israel where they would have been among the most exotic groups of Jews, both in physique and in traditional dress.

ABO BLOOD GROUPS

As in the case of Jews of all other regions, we have much fuller information on the ABO blood groups of the oriental Jews than on their blood groups of other systems. As shown in Fig. 2, there is a wide scatter of frequencies, but there is an

even wider one for the ABO groups of the populations among whom they have lived.

There is, as already mentioned (p. 17), no possibility of finding in Israel any sample representative of the original pre-Diaspora Jews. The tested samples of unclassified Israel-born Jews can only represent a weighted average of immigrant populations from all over the world, and the close resemblance of their blood group frequencies to those of the Palestine Arabs is presumably purely accidental.

The Babylonian Jews from Iraq may possibly give a clue to the genetic composition of the Jews of the early forced emigrations, and it may be meaningful that their ABO gene frequencies lie not far from the centre of the cluster of Oriental Jewish frequencies, especially as the Babylonian Jews differ considerably from the present Arab population of Iraq, having higher frequencies of both the *A* and the *B* genes. Nearly every other Oriental Jewish community differs in this manner from its non-Jewish neighbours, and the same is true of a large proportion of Jewish communities outside Asia. It is also, however, generally true that fluctuations in *A* and *B* frequencies in indigenous populations are accompanied by corresponding fluctuations, though around higher average levels, in the Jews who have lived among them.

The generally high *A* frequencies, and to an even greater degree the high *B* frequencies, of the Jews present a general problem. In the case of Africa the suggestion is made below (p. 37) that the Jews, on their exodus into Palestine and their re-entry into Africa, had preserved something of the high *A* and *B* frequencies of an Egyptian portion of their ancestry.

This could apply also to the Jews of Asia, but here the *A* and *B* frequencies are on the whole higher than those of the African Jews. Thus, if we are to account for the high *A* and *B* frequencies mainly or largely in terms of hybridization, we must assume mixing with peoples from farther east in eastern Iran and central Asia.

Two small groups, the Karaites of Hit in Iraq and the Bene-Israel Jews of Bombay, claiming to be offshoots of the Babylonian Jews, have particularly high *B* frequencies. These Karaites, however, who must have separated from the parent population over a thousand years ago, present a special case, for they have 23 per cent of *A* and the exceptionally high frequency of 35 per cent of *B* genes; most of their other blood-group frequencies also are exceptional, and such that they could not have originated from any conceivable set of hybridizations alone. It must therefore be supposed that because of their small numbers they, like the Samaritans, have been affected to a high degree by genetic drift.

The so-called Baghdadi Jews of Bombay have 25 per cent of *A* and 28 per cent of *B* genes. As already mentioned (p. 25), they came from the Near East as merchants in the early nineteenth century, almost too recently to be greatly affected by genetic drift. Their ABO frequencies could, arithmetically, be explained by hybridization with Indian or perhaps Afghan peoples, but they have particularly pale skins, and undoubtedly avoid marrying indigenous Indians, for they even refrain from intermarrying with the Bene-Israel Jews of the Bombay area who are markedly darker than themselves.

The Bene-Israel Jews do, indeed, have ABO frequencies almost identical with those of the local Indians with whom they have certainly intermarried.

There is some doubt as to the precise ABO frequencies of the Cochin Jews of Kerala. Not only do the 'White Jews' have higher *A* frequencies than the 'Black Jews', but there are discrepancies between results of tests done in India and in Israel. However, average frequencies of 12 per cent of *A* and 14 per cent of *B* genes are very close to frequencies found in the indigenous Malayali. Thus there has almost certainly been intermarriage between this ancient Jewish community and the indigenous population.

OTHER BLOOD GROUP SYSTEMS

For the systems other than ABO only a few of the populations just mentioned have been tested. Since several of the remainder are represented in Israel it is much to be desired that additional tests should be done. Such tests would undoubtedly contribute materially to the unravelling of the many strands in the complex web of the history of the 'lost tribes'. Even where tests have been done, the numbers tested are small and the results are certainly affected by sampling errors.

Apart from the circumscribed high frequencies of the *M* gene in both Arabs and Jews of the Arabian peninsula, frequencies of this gene in indigenous populations of south-west Asia fall into a very narrow range, national averages lying between 56 and 63 per cent, with a general rise from west to east. The few data which we possess on Jews from this region conform to the same pattern, but are 2 to 4 per cent lower than in the corresponding indigenous populations, as shown in Table III.

The *M* frequency of 56·3 per cent found in the Babylonian Jews is remarkably close to the national average for indigenous Lebanese populations (56·5 per cent) and to the

average for Palestinian Arabs (56·6 per cent). Both *B* and *M* frequencies tend to increase eastwards in south-west Asia; thus for the *B* gene the Jews show a more easterly trend than their non-Jewish neighbours, while for *M* they show a more westerly one.

TABLE III. *M* GENE PERCENTAGES IN JEWISH AND INDIGENOUS POPULATIONS IN SOUTH-WEST ASIA

Iraq (Babylonian) Jews	56·3
Iraq Arabs	60·5
Iraq Kurdish Jews	53·2
Iraq Kurds	54·9
Iran Kurdish Jews	57·6
Iran Kurds	61·9

The Rhesus blood groups have been the subject of numerous surveys; the results show no clear trends though, apart from the highly aberrant Karaites, the Oriental Jews fit well into the south-west Asiatic regional range of variation.

For the Babylonian Jews of Iraq we have several consistent sets of data showing approximately the following percentage frequencies: *CDe*, 53; *cDE* 16; *cDe*, 4; *cde*, 24. These figures may conceivably represent something like the composition of the original Jews and, because of the importance of Mesopotamia as a source of dispersal into Europe, they represent a valuable baseline. The *cDe* frequency, though low, allows the possibility of inclusion of some African genes. The *cDE* frequency is somewhat above the usual Mediterranean levels and suggests some incorporation of genes from the north, i.e. Kurdistan and beyond.

The different sets of data on Jews from the Iraq portion of Kurdistan are rather inconsistent with one another, perhaps because the Kurdish Jews were a set of isolates. However they all show approximately 20 per cent of *cDE*, or considerably above the levels found, on the basis of very small numbers, for the non-Jewish Kurds of Iraq and Turkey. The frequencies found for *cDe* in the Iraqi Kurdish Jews show considerable variation, from 0 to 15 per cent, within the same range as those found for the Kurds of Iraq and Turkey. The variation found is too great to permit any definite statement on the general degree of African admixture either in Kurds or in Kurdish Jews, but the zero frequency found by Tills *et al.* (1977) in two separate small groups of Jews suggests the possibility that there was little or no African admixture in Jews of the original dispersion into Assyria under Sargon.

The Jews of Iran, presumably an offshoot of those of Babylonia, were scattered in many parts of the country, which probably accounts for the variability of their Rh blood-group frequencies. They have, however, on the whole a high frequency of *CDe* and a low one of *cDE* as compared with Babylonian Jews as well as with non-Jewish Iranians. The observed difference may, however, be a result of sampling error. The frequency of *cDe*, averaging about 4 per cent, is low like that in the Babylonian Jews and suggests that there has been little intermarriage with populations of African ancestry.

The Kurdish Jews of Iran are very close in their Rh frequencies to the Kurdish Jews of Iraq, but they have a high *cDE* frequency. They also closely resemble the Iranian Kurds.

The Bene-Israel and Baghdadi Jews of Bombay, with 34 and 31 per cent respectively of *d* genes, fit well into the Indian Rh picture, though they have rather more of these genes than most Oriental Jewish populations. However, the Jews of Cochin, Kerala, have 44·4 per cent of *cde* and 2·4 per cent of *Cde*. The total of 46·8 per cent of *d* genes is quite exceptional for any Jewish or Indian populations and even above the level of most non-Jewish west European peoples.

The non-Kurdish Jews both of Iraq and Iran have high frequencies of the Kell *K* gene, comparable to those found in the Arabs of Arabia. The Kurdish Jews of Iraq and Iran also have several per cent of the gene, as have the non-Jewish Kurds both of Iraq and Iran, as well as most Iranian populations. This is mainly a European gene with low frequencies in most of Asia, apart from this south-western corner.

The Duffy *Fy⁴* gene is recorded as having a frequency of about 15 per cent in Kurdish Jews both of Iraq and Iran, as well as in Iranian Kurds. While this is essentially an African marker gene, with frequencies above 90 per cent in most Negro populations, its frequencies in the north-east corner of Africa, and in south-west Asia suggest that it may have a substantial frequency in this area in peoples with only a small amount of recent African ancestry. This, as discussed elsewhere (p. 9). may be related to its protective effect against vivax malaria.

Because we have so few comparative data, the plasma-protein and red-cell enzyme factors, with one exception, tell us little about the relationships of the Jewish and non-Jewish populations of south-west Asia. This one exception, glucose-6-phosphate dehydrogenase (G6PD), has become proverbial. Nearly all Jewish populations have substantial frequencies of the *Gdᴮ⁻* gene which confers a deficiency of enzymic activity. The deficient type is almost absent among the Ashkenazim, but the highest frequencies of it known from any population, Jewish or gentile, are found in the Kurdish Jews, especially those of Iraq.

Deficiency of G6PD, which may be due to either of the genes *Gdᴮ⁻* or *Gdᴬ⁻*, or to certain other rarer variants, tends, as already explained (p. 9), to cause haemolytic anaemia on consumption of certain foods and drugs. On the other hand it confers some protection from malaria. The next highest known frequencies, after those of the Kurdish Jews, are found in the Shia Muslim Arabs of the Qatif Oasis of Saudi Arabia, a formerly highly malarious area. Here, as in many other regions, the raised frequency of G6PD deficiency can be explained as a result of natural selection, the benefits of protection from malaria being compensated by deaths from haemolytic anaemia, so that when different regions are compared, there is a correlation between the present or former incidence of malaria and the frequency of G6PD deficiency.

This relationship appears to account, over very large areas, including Saudi Arabia, for the incidence of the enzyme deficiency, but it does not by iself explain the exceptionally high frequency in the Kurdish Jews. There is no evidence that malaria was exceptionally rife in the regions, mostly mountainous, where they lived. Moreover, their dwellings seem to have been closely adjacent to those of the non-Jewish Kurds who have very much lower frequencies of G6PD deficiency, the *Gdᴮ⁻* gene frequency (and the phenotype frequency in males) being only 4 per cent in non-Jewish Kurds from Syria and Turkey (combined sample) and in those of Iran (there seem to be no data for those of Iraq), compared with figures of up to 70 per cent in Kurdish Jews from Iraq, and 37 per cent in those from Iran.

Frequencies in Jews from Baghdad, and from the Caucasus, are 24 and 28 per cent respectively, high levels but matched by those in numbers of non-Jewish populations.

The deficiency gene may have been carried into Assyria by the original deported populations from Israel, especially those coming from the malarious swamps surrounding the former Lake Huleh in the extreme north of modern Israel, but the very high frequencies now found can be explained only by a continuing and severe process of natural selection, possibly related to environmental factors of a cultural nature.

The possible kind of process is suggested by the example of the Parsis of Bombay. Their rather high level of G6PD deficiency, compared with that of the Indians among whom they live, was explained by Undevia (1969). In conformity with their relative wealth, and their high standard of living and of hygiene, the Parsis live in compounds of inwardly-facing buildings with an open space and an open water tank in the centre. This formerly promoted the breeding of mosquitoes which in turn caused malaria and, by natural selection, a high incidence of G6PD deficiency.

The phenylthiocarbamide taster frequencies of most Oriental Jewish populations do not differ greatly from those of the surrounding peoples. Again, however, the Indian Jews of Cochin, Kerala, stand out with the unusually low frequency of 35 per cent of the taster gene. This is below the levels found in most other Jewish populations. It is also below the average Indian level, though matched by a number of indigenous populations in Madras and Kerala as well as elsewhere.

The genetics of the Jews of Asia will be discussed further when we consider the Ashkenazim, and the extent to which Asiatic Jews may have contributed to the gene pool of eastern European Jews.

6

THE YEMENITE JEWS

SOME Jews may have migrated to the Yemen at the time of the first Diaspora under Nebuchadnezzar, but it appears certain that the largest migration to that country came at the time of the crushing of the two great Jewish revolts, in A.D. 70 and 132–5. The main Jewish migration into Arabia was undoubtedly to Himyar or Yemen, but Jews settled also in many other parts of the Arabian peninsula, especially near Medina. Within the next two centuries they acquired great influence in Himyar as well as further north. There must have been numerous conversions of the hitherto pagan Arabs to the Jewish faith, but we cannot, solely from historical records, tell how extensive these were. Blood-group data, however (pp. 31–2) suggest that the conversions were extremely numerous. Though the record is not continuous, the Yemen or Himyar appears to have constituted a kingdom mainly under Jewish rule from about A.D. 200 to 460 when the ruler, 'Abd Kulalem, embraced Christianity, though at least two subsequent rulers followed the Jewish religion.

The pagan Arabs seem to have offered little resistance to the spread of Judaism, and the chief enemies of the Jewish Himyarite rulers were the Christian Ethiopian rulers based on Axum just across the southern Red Sea.

The advent of Islam in the early years of the seventh century A.D. marked the end of Jewish dominance in the Yemen. Many Jews were undoubtedly forcibly converted and probably many killed, especially in the north where their political power and control of trade were completely abolished. However, some Jewish scholars were attracted by the new religion and voluntarily migrated to Medina where their knowledge of Jewish lore made a valuable contribution to the developing traditions and literature of Islam.

In the south a very considerable body of Jews maintained their faith and became the founders and ancestors of the Yemenite Jewish community, which has now migrated to Israel. It is clear that Mohammed was much less insistent on conversion or elimination of Jews in this region than of those living near Medina.

Throughout subsequent ages we have only occasional glimpses of the Yemenite Jews, from the records of travellers and historians such as Maimonides, Benjamin of Tudela (twelfth century), and Obadiah of Bertinoro (fifteenth century).

Despite persecution they not only maintained their Jewish faith, but, as is clear from the literature which they recently brought with them to Israel, they kept in touch with Jewish thought and literature, and also produced able writers of their own. They made their living chiefly as skilled artisans, and their work as silversmiths is now famous. This is a surprising development, for most Jewish populations have avoided the plastic arts, as contravening the prohibition of making 'graven images' (but see also p. 37).

On the south-eastern border of the Yemen, at Beida, and at Habban in the Hadhramaut, a group of Jews, now known as the Habbanite Jews, was cut off almost completely from contact not only with the world at large but with other groups of Jews. They however maintained their religion and their communal identity until, in 1950, all 345 then living emigrated to Israel.

From time to time in pre-Islamic days Jewish-ruled principalities emerged also in other parts of Arabia. One of these was that of Khaibar, about 140 km NNW of Medina. The Jewish inhabitants of the city of Khaibar, like those of Medina, were ruthlessly exterminated by the early Muslims, but in the surrounding desert nomad tribes of Jewish religion who called themselves Rechabites—followers of the ascetic practices of the Sons of Rechab (Jeremiah, ch. 35)—continued in existence until the present century. Ben-Zvi (1958) relates how Abu'Ish, Mukhtar or leader of an Arab village in southern Palestine, whom he describes as 'the last of the tribe' was executed in the city square of Gaza during the Jewish war of independence, on suspicion of being a Jew and a spy.

As we have seen, the main body of Yemenite Jews retained contact with world Judaism, and in the days of emergent Zionism they were the first, in the 1880s, to migrate in substantial numbers to Palestine. Already, before the foundation of the State of Israel the migrant community numbered 35 000, and subsequently the whole of those remaining in the Yemen moved to Israel.

In the days of the British Colony a Jewish community of about 4000 persons lived in the Old City of Aden ('the Crater'). They were engaged mainly in trade, and one may surmise that their origins were somewhat cosmopolitan, though no doubt derived partly from the Yemen communities. A large number, and perhaps by now all of them, have migrated to Israel.

BLOOD GROUPS

The Yemenite Jews constitute one of the largest ethnic communities in Israel, and they have been subjected to a great variety of scientific investigations, including serological ones. Some studies, especially those for the ABO blood groups, have involved very large numbers of persons. Most of these have not made any distinction between those coming from different parts of Yemen, but an intensive study (Tills et al. 1977) had distinguished between those from the south and those from the north of the Yemen Republic. Because of the small numbers tested, differences between the immigrants from these two areas do not, in the case of most systems, reach a statistically significant level, but such differences as have been found are of some interest.

There is a fairly good agreement between all the numerous sets of data on the frequencies of the ABO groups. The overall frequencies of the A, B, and O genes are 19·5, 8·7, and 71·8 per cent respectively.

These frequencies fall close to those found in the Yemenite Arabs (18, 7, and 76 per cent, on rather small numbers) and in the Zabidi Arabs of the Yemen (17, 8, and 75 per cent). Other neighbouring populations with similar A and B frequencies are the Bedouin of Saudi Arabia (13, 12, and 72 per cent) and, rather surprisingly, the Samaritans (17, 7, and 76 per cent). The overall *A*, *B*, and *O* gene frequencies of the Saudi Arabians, based on a very large sample, are 13·4, 12·4, and 74·2 per cent respectively. Most other populations of the Near East have distinctly higher *B* and lower *O* frequencies. The mean frequency of the A_2 gene in the Yemenite Jews is very high, the overall level being 8·9 per cent, very close to the 9·6 per cent found in the Samaritans. These values are considerably higher than the 4·9 per cent found in Saudi Arabia (where however the Bedouin have 7·5 per cent). The Southern Arabs, also, have a high frequency of A_2—8·9 per cent (together with 2·4 per cent of the *A*-intermediate gene).

The Yemenite Jews have the very high *M* gene frequency of 77 per cent, including 36 per cent of the *MS* complex. The Habbanite Jews show closely similar frequencies. These high levels are matched by those of the Arabs of the Arabian peninsula; the Bedouin of Saudi Arabia and the Arabs of southern Arabia show almost identical frequencies. The frequencies found for the Yemenite Arabs are slightly lower (*M*, 74 per cent; *MS*, 31 per cent) but these are based on a small sample of only 104 individuals.

The Henshaw African marker gene has an overall frequency of 2 per cent in the 179 Yemenite Jews tested for it, but all 7 examples of the antigen were in the 75 persons from the southern part of the Yemen Republic. None were found in the Habbanite Jews. The overall gene frequency in the Saudi Arabs is 0·4 per cent and in the southern Arabs 0·2 per cent.

For the Yemenite Jews and neighbouring non-Jewish populations the Rhesus system shows more erratic gene frequencies than ABO or MNS, probably because of sampling errors arising from the large number of alleles under consideration, in proportion to the relatively small numbers of individuals tested. It is thus almost impossible to see any regularity in the frequencies of the *CDe*, *cDE*, and *cde* complexes.

There is, however, rather more regularity in the frequencies of the African marker genes and gene complexes. If we add together all the variants of R_0 or *cDe*, i.e. *cDe*, $cD^u e$ and the combinations of these two with the *V* gene (where appropriate tests have been done), the total R_0 frequencies in the northern and southern Yemenite Jews, and the Habbanite Jews are 2, 10, and 29 per cent respectively, while the frequencies of the *V* gene in the same three populations are respectively 2, 7, and 23 per cent. In both these respects the Yemenite Jewish frequencies are near the lower limit of those found in neighbouring Arab populations, and those of the Habbanite Jews higher than in any of the Arab populations, with the exception of the Jebeliya Bedouin of the Sinai Peninsula.

The Kell antigen K reaches some of its highest known frequencies among Arab populations; the highest *K* gene frequencies among them are 10·56 per cent in the Bedouin of Saudi Arabia and 18 per cent in the Jebeliya Bedouin. In this respect, however, the Yemenite Jews differ markedly from the Arabs, having much lower gene frequencies, near 1 per cent. The closely linked Js^a gene, which is an African marker, has a frequency of 1·3 per cent in the southern Arabs and 6·5 per cent in the Jebeliya; much higher frequencies, up to 25 per cent, are found in some African populations. In the northern and southern Yemenite Jews and the Habbanites gene frequencies are respectively 0, 1·1, and 6·7 per cent.

The Duffy system also includes a gene, Fy^4, which is an African marker. The very high frequencies found in Near Eastern populations, both Jewish and Arab, are however, surprising. Some of the recorded frequencies may be unduly high because of false negative serological results, due to the use of very weak anti-Fy^b reagents—unfortunately specific anti-Fy^4 serum is in extremely short supply. It has however been suggested by Miller (1975) that $Fy^4 Fy^4$ homozygotes are protected against malarial infection by *Plasmodium vivax*. Thus the frequency of the Fy^4 gene, even outside Africa, may in the past have been raised by natural selection in malarial environments. Owing to the very few tests done with anti-Fy^b, Fy^4 gene frequencies are subject to considerable uncertainty, but estimates of 61, 82, and 58 per cent have been made for the north and south Yemenite Jews and the Habbanite Jews respectively.

The Jk^a (Kidd) gene, which reaches higher levels in African than in Caucasoid populations, has frequencies of 57, 69, and 63 per cent respectively in the three populations just mentioned, as compared with averages of 50 per cent in Europeans and 67 per cent in African Negroes.

The Diego (Di^a) Mongoloid marker gene is present in the Hadhramaut Arabs but has never been recorded in any Jewish population in the Old World, though it was probably present in the Chinese Jews of K'ai-fêng Fu, and it should be sought in Bukhara Jews. (It is in fact found in the Sephardim ('Spanish' Jews) of Argentina, who presumably acquired it from an American Indian source.)

Other polymorphic systems do not yield much information that is of use in studying the relationships of the Yemenite Jews. In the haptoglobin system they show Hp^1 gene frequencies just below the moderate European levels, but rather above the low ones typical of Asia, and very much below the high African levels.

The Habbanite Jews are remarkable for their exceptionally high frequency, 58 per cent, of the phosphogluco-mutase allele PGM_1^2. The frequency of 42 per cent in the northern Yemenite Jews is also above general world levels which average about 25 per cent.

In the acid phosphatase system the frequencies of P^a are low in northern Yemenite and Habbanite Jews, as they are in most African populations; P^c, which is rare or absent in Africa, has frequencies of 9 and 7 per cent in north and south Yemenite Jews, but only 2·7 per cent in Habbanites.

Frequencies of the 6-phosphogluconate dehydrogenase gene, PGD^c, are near African levels, and above the low levels characteristic of Europe and Asia.

The levels of glucose-6-phosphate dehydrogenase deficiency are of particular interest in Jewish populations, in nearly all of which it is present except for the Ashkenazim (Jews of eastern Europe). The frequency of deficient genes (probably mostly Gd^{B-}) in Yemenite Jews is low, being about 3 per cent in both north and south Yemenite Jews, and 1·3 per cent in Habbanite Jews. In Saudi Arabs its frequency is mostly below 5 per cent, but reaches very high levels in the malarious oases of the north-east. These high levels are further discussed in relation to the similarly high levels in the Kurdish Jews (p. 29).

The blood-group frequencies of the Yemenite Jews define their genetic relationship to surrounding populations in a more precise way than is the case for any other set of Jews. They are very closely similar to neighbouring Arab populations—slightly nearer to those of southern Arabia than to those of Saudi Arabia. They are also very near to the Arabs of the Yemen itself, but we have less data for the Yemenite Arabs than for those of the other parts of Arabia. However, besides their main set of Arab-like genes, the Jews have a well-defined set of African marker genes which enable us to say that both the Yemenite Jewish populations have some

African genes, but these are distinctly commoner in the south than in the north. In the Habbanite Jews they are higher still and comparable with those in the Arabs, but frequencies are rather erratic, almost certainly because of genetic drift in this small isolate. The Habbanites bear some resemblance, in their high level of African genes and in the erratic nature of their general blood-group picture, to the Jebeliya of the Sinai Peninsula. It is not of course suggested that there is any direct relationship between these two isolates, but merely that they had broadly similar histories. However, we have a very good idea as to why the Jebeliya incorporated so many African genes—it was almost certainly because the early Muslims would not exchange women with them on account of their association with a Christian monastery. One wonders if some similar cultural influence was at work in the case of the Habbanites.

It is thus clear that, broadly speaking, both Jews and Arabs of southern Arabia are descended from the population of the pre-Muslim Himyarite kingdom, many of the rulers, and probably also the subjects, of which followed the Jewish religion. It would appear that the religious segregation into Jews and Muslims at the time of the Muslim conquest was mainly a cultural rather than a genetic one, but we cannot tell how completely the pre-Muslim population was Judaized, or what proportion of those who accepted Islam had previously been Jews. As we have already seen, in northern Arabia there was much enforced conversion, but this seems to have been much less the case in the south.

This however still leaves open the question of the origin of the essentially uniform pre-Muslim population of Arabia, which, on the evidence of its modern Arab and Jewish descendants, differed to a high degree from all surrounding populations. Probably this was an autochthonous population, that is to say, one which had evolved over a period of hundreds if not thousands of years in this relatively inaccessible sub-continent. The Jewish religion must however have been introduced by migrants, probably at the time of the Jewish revolts early in the Christian era, and these migrants may have been sufficiently numerous to modify the overall genetic composition of the population substantially, if indeed the two peoples did initially differ at all considerably in a genetic sense. The Samaritans, greatly altered as they must be by genetic drift, are nevertheless probably 'purer' descendants of the Jews of two thousand years ago than any other living community, and it is therefore interesting to see that they do bear a certain resemblance to the Yemenite Jews.

A final word must be said about the Jews of Aden. The few data which we have about them show them to be very different from the Yemenite Jews, and they are likely to be, as already suggested, of somewhat cosmopolitan origin, but further studies might make it possible to define their origins more precisely.

7

THE KARAITES

THE Karaites are the 'Protestants' of Judaism, a sect which has accepted the Bible (the Old Testament) but rejected the later interpretative literature and traditions incorporated in the Mishnah and the Talmud, accepted by most other Jewish communities. Their own interpretations have however been highly restrictive, though in a different way from the traditional ones which they have claimed to supersede.

The movement was founded in Babylon about the middle of the eighth century A.D. by a leader of the Jewish community there, Anan-ben-David (Epstein 1959).

At first the movement spread widely, establishing itself not only in Egypt but also as far away as Morocco and Spain, though the community set up by the founder himself in Palestine did not long persist.

In the writings produced by the community in Egypt a new influence shows itself in the ninth century (Allegro 1961). This was apparently derived from documents said to have been found in a cave near Jericho, and almost certainly originating from the Qumran sect (q.v.).

The Karaite movement stimulated a deeper study of the Bible and a flourishing of Talmudic studies among the orthodox. By about the middle of the tenth century the Karaite movement had almost ceased to spread, but important Karaite communities have, nevertheless, persisted to the present century.

Two of the original Karaite communities remained in existence until they were able, in recent years, to move to Israel. One of these was the large Egyptian one. The other was the Babylonian one, latterly living at Hit (Iraq). It possessed an old synagogue which was destroyed just before the Arab–Israeli war of 1948, and an ancient scroll of the Law, which was seized by the Iraqi government authorities.

Just before the Second World War the majority of Karaites lived in the Soviet Union and what was then eastern Poland; in addition to the Egyptian and Babylonian communities mentioned above there was a small one at Istanbul (Turkey) and apparently another in Tunisia (since six Karaite families are recorded as coming to Israel from Tunisia in 1948).

In the 1930s a group of Italian scholars visited the Karaite communities in eastern Europe. They included a Dr. Gil who later became Deputy Director of the Department of Statistics of the Government of Israel. Ben-Zvi (1958) refers to their observations as follows:

> The Crimean Karaites betrayed a high admixture of Jewish and Tartar physiognomic traits; the Karaites on the Volga, who betrayed a decisive Slavic admixture, were probably proselytes of Russian descent. The only Karaites who had no Russian but did have definite Tartar and Khazar traits, were those of Poland who had migrated in the fourteenth century from their native Crimea. The explorers reached the conclusion that some of the Karaites in Eastern European countries were originally people of non-Semitic stock who adopted the Jewish faith, and in course of time mixed and intermarried with the Karaites of Jewish racial descent.

On the other hand Ben-Zvi states that the Karaites of the Near East were physically of Semitic appearance.

The Karaites of eastern Europe made considerable use of Hebrew but in everyday life spoke a Tartar dialect, thus showing that at least the nuclei of the communities came from the Crimea.

During the Second World War the Karaites, in the areas overrun by the Germans, claimed to be Jews only by religion and not by 'race'. This claim was accepted by the German authorities and the Karaites thus escaped the extermination suffered by the orthodox Jews. This distinction was made in the Crimea as well as in the rest of eastern Europe; blood-group data were already available at this time, and it is thought that they may have been used as part of the basis for discrimination.

Indeed, as early as 1783, when the Russians first seized control of the Crimean peninsula from the Tatars, the Karaites of that area were already a community quite distinct from the Krimchaks (q.v.) or orthodox Jews of the same area, and were exempted from the restrictions applying to the latter.

Few, if any, of the European Karaites have migrated to Israel, but about half of the large Egyptian community has done so, as well as almost the whole of the small one of Hit in Iraq.

BLOOD GROUPS

It is generally admitted that the European Karaite stock, whatever may have been its origin and original Jewish component, now contains a very considerable and probably heterogeneous admixture of Crimean and other European contributions. The Iraq Karaites have long been strictly endogamous but because of the small size of the community are likely to have been subject to a considerable amount of genetic drift. The Egyptian community alone might have given us a genetic clue to its origins, but it has been tested only for A_1A_2BO.

It is nevertheless interesting that all the four communities tested show unusually high frequencies of the B gene when compared with surrounding indigenous populations, and with some though not all neighbouring orthodox Jewish ones.

The Egyptian Karaites have the extremely high B gene frequency of 34·5 per cent, very much higher than that of other Egyptian Jews, and than that of indigenous Egyptians (who themselves have more B genes than neighbouring indigenous populations). The A gene frequency of the Egyptian Karaites, on the other hand, is 7·2 per cent, very much lower than the levels, near 25 per cent, found both in orthodox Egyptian Jews and in indigenous Egyptians. The values found are suggestive of considerable genetic drift despite the large size of the community. It would be of great interest to carry out tests for several other genetic systems on the considerable numbers now living in Israel.

The Karaites of Hit, Iraq, have a slightly higher B gene

frequency even than those of Egypt, but differ in having 22·5 per cent of A genes, which is near the normal levels of many neighbouring Jewish and indigenous populations.

The Karaites of the Crimea have 21 per cent of A and 18 per cent of B genes, figures which are slightly lower and higher respectively than those of the indigenous populations of the region. The orthodox Krimchaks (q.v.) of the area, now largely exterminated, had considerably higher frequencies of both A and B. The Karaites of Lithuania have 11·5 per cent of A and 24 per cent of B genes, which are considerably lower and higher respectively than the indigenous Lithuanians and other adjacent peoples, including the orthodox Lithuanian Jews. It is interesting to see that the B frequency is near to that of the Krimchaks (but not of the Karaites) of the Crimea, but the Krimchaks had a much higher A frequency.

MN frequencies are known only for the Karaites of Lithuania and of Iraq. The M gene frequency of those of Hit, Iraq, 98 per cent, can be due only to an extreme degree of genetic drift. The rather high level, of almost 70 per cent, found in the Lithuanian Karaites, like their high B frequency, suggests derivation of genes from the East. The blood groups and origins of the Karaites of Lithuania and the Crimea are further discussed on p. 52.

In the Rh system the Karaites of Iraq have 53 per cent of the Rh-negative cde complex, one of the highest levels known, comparable, in the Near East, only with that found in the Sinai Bedouin. They also have 21 per cent of the African marker gene complex, cDe, one of the highest frequencies recorded in a Jewish population. However, since their exceptional M and cde frequencies are almost certainly due to genetic drift, that of cDe may be due to the same cause and not to an exceptional amount of African admixture.

They also have the rather high K (Kell) gene frequency of 9·5 per cent. This however tends to be high in the region, both in Jews and Arabs.

The Fy^a gene frequency of the Iraq Karaites is not exceptional for the region. It is not low enough to suggest any large frequency of the Fy^4 African marker gene found in neighbouring Arab and Jewish populations.

The Iraq Karaites show normal regional levels for variants of haptoglobin, phosphoglucomutase, and adenylate kinase. Eighteen Iraqi and 250 Egyptian Karaites were tested for deficiency of glucose-6-phosphate dehydrogenase, but none were deficient. This deficiency is found, in varying proportions in nearly all other Jewish populations of the Near East.

The data at present available on the Karaites of Egypt and Iraq do not throw much light on their origins but, as will be seen (p. 52), those of the European Karaites have highly suggestive implications.

8

THE JEWS OF AFRICA

THERE are, or were until only a few years ago, substantial Jewish populations in Egypt and in all the countries of north Africa from Libya to Morocco, and the Falashas, or 'Black Jews', are still a very numerous people in Ethiopia. Also, apart from peoples whose religion, even if it is not orthodox Judaism, is essentially Judaic, there are in Africa many others, essentially Muslim or pagan, who follow a number of Jewish practices.

The first arrival of Jews in the countries of north Africa other than Egypt took place early in the Christian era, or possibly long before that, and the descendants of such early immigrants formed, and possibly still form, a majority of the Jewish population. However, in many of the coastal towns, where most blood-grouping tests have been done, most of the recent Jewish population consisted of Sephardim, descended from the Jews who, with the Muslims, were expelled from Spain following the final Christian victory of A.D. 1492.

When the biology of the modern complex Jewish populations of north Africa is examined, the Sephardic component will of necessity be examined together with the more ancient components. However, the history of the Sephardim as a whole will be considered in a separate chapter, and in the historical part of the present chapter we shall be concerned mainly with the earlier part of the north African story.

There are, of course, many Jews in South Africa, but these belong almost entirely to the European branches of Jewry, and will be considered together with the latter.

Egypt is, geographically speaking, very easily accessible from Palestine. It was almost certainly over the Isthmus of Suez that man first entered Eurasia, and that the presumably Caucasoid Capsians, going the other way, entered north Africa in mesolithic times and became the main ancestors of the Berbers about whom much will be said below.

Any hindrances to movement between Palestine and Egypt have thus been much more political than physical and, over long periods, Egypt has served as a place of refuge from Palestine, from the time of Jacob and his sons to that of Jesus and his parents.

The history of the Israelites as a nation, rather than as a small tribe, may be said to begin with the Exodus from Egypt, but the Exodus, by its very nature, left behind no sojourners. Subsequently, as the Israelites became established in Palestine, there must have been many small migrations to Egypt, but one of the first to give rise to an established Jewish community in Egypt was that which took place when Nebuchadnezzar took Jerusalem in about 587 B.C. At this time considerable numbers of people from the land of Judah fled to Egypt, and among them was the prophet Jeremiah.

However, as shown below, there are indications that, long before this, colonies of Israelites had migrated by sea to more distant parts of north Africa.

Later, the conquests of Alexander the Great converted the eastern Mediterranean into a Greek lake; at his death in 323 B.C. his lands were divided between his generals; at first the Ptolemies of Egypt also ruled Palestine and Phoenicia.

They encouraged Jews to settle in Egypt and especially Alexandria, and it was here that the Old Testament was translated into Greek, some time before 200 B.C. According to tradition this was done by seventy-two scholars, hence the name 'Septuagint'. From Egypt the Jews spread into Cyrenaica, and thence probably further west still. Meanwhile Rome had become the ruling power in the east Mediterranean, though Greek remained the lingua franca. A rebellion of the Jews in Cyrenaica in A.D. 117 was ruthlessly suppressed by the Emperor Trajan. This led to migrations to the west where the power of Rome was less pervasive.

There are, however, strong indications that long before the Christian Era, and even long before the time of Nebuchadnezzar, there were Israelites in many of the countries of north Africa.

Most of the evidence for this comes from the work of Professor N. Slouschz (1927), linguist, historian, epigrapher, and traveller, who, during the early years of the present century, visited almost every Jewish community of any size in the whole of north Africa, from Cyrenaica to Morocco, suffering great hardship and being several times in danger of his life from bandits and from disease. He deserves a place of honour alongside the great Jewish travelling historians who, by visiting outlying Jewish communities, did so much during the Middle Ages to maintain the doctrinal and literary unity of Judaism and to transmit a knowledge of these communities to the rest of the Jewish world and to future generations.

The Phoenicians, from Tyre and Sidon, in the land which is now Lebanon, established colonies on much of the coast of north Africa about the time of David and Solomon, the latter of whom was a friend of Hiram King of Tyre who supplied cedar wood for the Temple. The most important of their colonies was Carthage, near the modern city of Tunis.

Slouschz puts forward two main lines of evidence to support his claim that Israelites accompanied the Phoenicians. One is that the language of the numerous inscriptions left by the colonists is, both in vocabulary and in alphabet, almost identical with contemporary Hebrew. He mentions, without precise details, that colonists included numbers of members of the northern maritime tribes of Zebulon and Asher. Whether, however, the Phoenician colonists were accompanied by racial Israelites or not, there is evidence of a great spread of Palestinian tradition at this time among the Berber tribes of north Africa, so that many if not most of them now believe that they are descended from the Philistines, the inhabitants of south-western Canaan, driven out by the early Israelites. They all know the story of the defeat of Goliath the Philistine by David. A great many other tribes, now Muslims, follow a number of Jewish customs, and have traditions that they were formerly Jewish. Some of these traditions may however refer to the later proselytization mentioned below.

The evidence of written history, and of inscriptions, shows the presence all over north Africa of Jewish communities in the early years of the Christian Era. They were persecuted by the Byzantine Christians but better treated by the Vandals who succeeded the Byzantines in the sixth century A.D. It is likely that there was at this time much proselytization of Berber tribes to the Jewish religion. Then, at the time of the great invasion in the seventh century, by racially mixed but Arabic-speaking Muslims, the Jews united with the Berbers to repel the invaders. The best known story is of Dahyah al-Kahina, Jewish Queen and Priestess, who seems to have combined many of the characteristics of Sisera, of Boudicca (Boadicea) and of Jeanne d'Arc. Commanding a combined army she for years held the invaders at bay until, deserted by her Berber and Christian allies, she fell in a final battle.

Following the Muslim victories throughout the whole of north Africa, there were massive conversions of both Berbers and Jews to Islam, but still a great many Jewish communities, especially in the mountain regions, maintained their faith, as evidenced by the findings of many travellers throughout the centuries, particular Slouschz in the early years of the present one.

It is moreover clear, both from what we know of the numbers of the invaders, and from recent observations on blood groups (Mourant et al. 1976a) and on other physical characters, that the Arabic-speaking Muslim invaders from outside Africa were insufficiently numerous to affect greatly the physical composition of the population as a whole.

When the Muslims in turn invaded and occupied Spain, many Jews from both Africa and Babylon followed them and, under the tolerant Muslim regime, there was a great flowering of Jewish scholarship in their new country. There is little doubt that considerable intermarriage of the original immigrant Jews with Spaniards took place, so that when the Muslims were finally driven from Spain in 1492, the Jews turned out with them must have carried both north African and Spanish genes, as well as some genes from the Babylonian community. Some of the expelled Jewish families, however, appear to have been of fairly pure aristocratic Spanish descent. The expelled Jews, henceforth to be known as the Sephardim, became dispersed throughout northern Africa and in many parts of southern Europe and the Near East. The genetical composition of the Jewish communities of north Africa was further complicated in the nineteenth and twentieth centuries, during the French regime, by new arrivals of Jews from Europe.

From the time of the Muslim invasions onwards the history of the north African Jews is fairly well documented, and we have a wealth of information on their situation during the present century, especially from Slouschz (1927) but also from Kossovitch (1953), Mechali et al. (1957), and Briggs and Guède (1964). It is unnecessary to give details here except about populations for which we have blood-group data, but a few general remarks should be made. Unlike most other Jewish populations (except in Kurdistan and modern Israel) many of the Jews in north Africa were engaged in agriculture. Many, however, like Jews elsewhere, were artisans, including metal workers. In this connection it is of particular interest to find that, especially in the south of the Maghrib, they included considerable numbers of blacksmiths. Further reference is made to this below (p. 37).

Among the communities practising agriculture there are many which have never been studied in detail, including the Jews of the Wadis Sous and Draa in south-western Morocco. These were the most westerly of all Old World Jewish communities, unless, as Slouschz suggests, there are Jewish groups even further south-west as far as Senegal, as well as in Niger to the south. Ben-Zvi (1958) quotes Slouschz,

without a precise reference, as claiming that, 'as late as the ninth and tenth centuries there was still an extensive Jewish kingdom that stretched from the hills of Ethiopia (see Ethiopia, below) through the vast expanses of the Sudan, and so far as the Atlantic coast, dominated by Jewish cavalrymen'.

Most of the blood-group tests on north African Jews have been done on emigrants, either in Israel or France, so that only in a few cases do we know precisely whence they came. A large proportion of these emigrants, and of persons recorded as tested in named large towns in Morocco, are likely to be Sephardim, which may largely account for the similarity between the relevant sets of data, discussed below.

Besides the Tafilalet Jews, mentioned below (p. 38), the only ancient communities for which we have blood-group data are those of the island of Djerba, the land of the lotus eaters, in the Gulf of Tunis. This fertile island has long been a place of refuge both for Jews and for heretical Muslim sects. The Jews have been settled there for many centuries and are almost certainly entirely pre-Sephardic. They mostly live in two towns or ghettos, Hara-Kbira (The Great Hara) and Hara-Srira (the Small Hara). Hara-Srira was entirely inhabited by Cohanim claiming descent from Aaron and Hara-Kbira by non-Aaronide families. Traditionally, the first Jews settled here as a refuge during the early Muslim invasions, while others are said to have fled here from Tripoli in the twelfth century A.D.

ETHIOPIA

If the earliest origins of the north African Jewish communities are wrapped in obscurity, those of the Falasha of Ethiopia are even more obscure. Tradition derives them from the Jews from the Court of King Solomon, who accompanied the Queen of Sheba back to Ethiopia, and it is from this legend that the modern Emperors derived their title of 'Lion of Judah'. Another hypothesis would derive then from the Jews of the Himyarite kingdom in Arabia, but a more likely one, supported by Ben-Zvi (1958), finds their origin in Jewish colonies in Upper Egypt, and hence indirectly in the great Jewish community which existed in Lower Egypt in the Hellenic period. The strongest evidence for this is that their Bible, now written in the Geez syllabary, which is the common vehicle of all the written languages of Ethiopia, is derived from the Septuagint.

It is at least certain that for hundreds of years during the Middle Ages there was a Jewish kingdom in the hill country of Ethiopia which fought with varying success against the Christian Amharas, and which produced a number of folk-heroes, notably Gid'eon who fell in battle in A.D. 1624. Their syncretic religion, though far from rabbinic Judaism, may be described as more Hebrew than pagan.

As in other parts of Africa, there exist also a number of ethnic groups with religion and customs which, though still farther from orthodox Judaism, incorporate a number of Jewish elements. One such group is that of the Qemant (Gamst 1969), now numbering about 20 000, who are illiterate peasants, subject to the Amhara, who follow an elaborate set of religious customs, including many Jewish elements, but more pagan than Jewish. As their only access to a literary culture is through their Christian Amhara overlords, they are gradually being absorbed by the latter.

RHODESIA

Perhaps even more remarkable, because of the remoteness

of their present home, far south of the Equator, are the
Lemba of Rhodesia (Hughes *et al.* 1976) an essentially
Negroid people, though some anthropologists claim that
they show some Caucasoid features. Their food laws are
essentially Jewish, they observe Saturday as the Sabbath,
they practise male circumcision and fairly strict endogamy
(with purification rituals for outsiders marrying into the
group). They have a tradition that they came from the north
and that their fathers did skilled metalwork for the Arabs.
This suggests possible connections with the modern Yemenite
and north African Jewish silversmiths, and possibly also
with the nomadic Jewish blacksmiths of Khaibar in Arabia,
and other Jewish blacksmiths on the northern borders of the
Sahara. The relationship of the latter to the other itinerant
smiths of the region, usually described as gypsy-like, needs
to be defined more precisely both culturally and anthro-
pologically. Somehow too, there must be some connection
with the beginnings of iron working which initiated the rapid
spread of Bantu-speaking tribes throughout most of southern
Africa.

BLOOD GROUPS

The greater part of the genetic data on the Jews of Africa
refers to the ABO groups, but there is enough information
on some other systems, especially MN and Rh, to define a
pattern of distribution, and to compare it with that shown
by the non-Jewish populations of the same countries. The
main difficulty in making such comparisons is that, as already
mentioned, most of the Jewish data refer to the large towns,
whereas the populations which are historically the most
interesting are those which are, or were, interspersed with
the more numerous Berbers and Arabs in the small towns,
villages and agricultural areas remote from the coast.

In Fig. 3 the *A* and *B* gene frequencies of Jews and non-
Jews are plotted on a graph in the manner described on pp. 3–
4. In the calculation of gene frequencies of the indigenous
populations, Europeans and Negroids have as far as possible
been excluded, so as to give as good a representation as
possible of the Berber (and Arab) stock. In the case of Egypt
only the population of Lower Egypt has been taken into
account.

It will at once be evident that there is a systematic pattern
present. In every case the Jews have a higher *B* gene frequency
than the non-Jews and, in all cases but one, a higher *A*
frequency.

With the exception of the Egyptians, the gene frequencies
for the non-Jewish populations cluster around *A*, 21 per
cent, *B*, 12 per cent. The high *A* and *B* frequencies of indigenous Egyptians have long been known to distinguish
them from all surrounding populations. This has been
discussed elsewhere (Mourant *et al.* 1976a) and probably
dates from before the period with which we are now concerned.

The Jewish populations, with the exception of that of
Libya, have gene frequencies clustering around *A*, 23 per
cent, *B*, 16 per cent, very close to the values found for the
non-Jewish Egyptians. The serological similarity between
these and the Egyptian Jews is striking.

Jews have almost certainly been present in large numbers
for a longer time in Egypt than elsewhere in north Africa,
and the observed frequencies could largely be explained if
Egyptian Jews had intermarried extensively over a long
period with indigenous Egyptians, and had then carried
Egyptian genes as well as their own original ones to countries

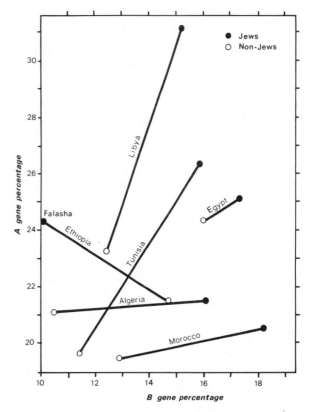

FIG. 3. Frequencies of the *A* and *B* genes in Jewish and indigenous
populations of north Africa.

to the west. This hypothesis does however leave some data
unexplained. An alternative suggestion, that the Israelites at
the time of the Exodus resembled the Egyptians in their
blood-group frequencies, somewhat increases the difficulty
of accounting for the genetic composition of non-African
Jewish populations.

For the Egyptian Karaites we have only gene frequencies
computed by Goldschmidt (1967), without the phenotype
data upon which they were based; the quite exceptional gene
frequencies, *A*, 7 per cent; *B*, 34·5 per cent, can hardly be
explained other than by genetic drift in a small endogamous
population. It is also impossible at present to explain the
very high *A* frequency of the Libyan Jews. Moreover, we
have not taken into account the unknown numbers of
Sephardim who are presumably included in the Jewish
populations tested, and who, when we meet them in Israel,
have an *A* gene frequency of only about 23 per cent.

Frequencies of the *M* gene must be seen against the background of the rather low *M* frequencies which characterize
all the non-Jewish peoples of north Africa except the
Egyptians, and which are especially seen in the Berber
isolates of the Atlas Mountains in Morocco. The overall *M*
gene frequencies vary from 46 per cent in Morocco to 52
per cent in Tunisia. That of the Egyptians is 56 per cent. In
every country concerned, except Algeria, for which there
are no Jewish MN data, the *M* frequency is higher in Jews
than in non-Jews. It reaches 58 per cent in Egyptian and
61 per cent in Moroccan Jews. It cannot thus be explained
entirely by *M* genes brought from Egypt and certainly not
by genes brought from Spain by the Sephardim. It may be
related to the extremely high *M* frequencies found in both

Jews and Arabs in Arabia; perhaps the original Palestinian Jews also had very high *M* frequencies.

The frequencies of the Rhesus blood groups in African Jews are, of necessity, more erratic because of the small numbers tested, particularly for antigens other than D. Overall frequencies of the *d* gene are near 34 per cent in all the countries concerned, except Egypt where the frequency is 29 per cent. None of these figures differ greatly from those found in Mediterranean populations in general. However, of particular interest are the frequencies of the R_0 or *cDe* complex. Nearly all Jewish populations everywhere show frequencies slightly above the usual European levels of 2 to 3 per cent, thus indicating a small amount of African admixture in the Jews. Frequencies in north African Jews vary from 6 per cent in Egypt to 10 per cent in Morocco, levels which are, in general, slightly higher than in the corresponding non-Jewish populations. The latter, indeed, show evidence of surprisingly little admixture of genes from the Negroes who must, for thousands of years, have faced them across the Sahara Desert, and who carry 60 per cent or more of *cDe*.

However the non-Jewish population of Lower Egypt has, according to two estimates, 17 or 23 per cent of this complex, either figure showing a very considerable African admixture. (The lower figure is derived from the data of Donegani *et al.* (1950) which were accidentally omitted from the tables of Mourant *et al.* (1976*a*).) It may thus be that the slight excess of *cDe* in north African Jews outside Egypt is due to some past admixture of Egyptian genes, but that part of the high level now found in Egypt is of recent origin and thus not reflected in those Jews whose ancestors long ago migrated westwards from Egypt.

Few other genetic systems supply data departing from the general Mediterranean pattern in such a way as to tell us much about the origin of these populations. The Libyan Jews appear to lack the African marker Duffy gene Fy^4, and where tests have been done only for Fy^a it is only in Morocco that the Fy^a gene frequency is reduced to a level suggesting the presence of several per cent of Fy^4.

The above accounts refer almost entirely to tests on what are probably rather mixed Jewish population samples, but tests have also been performed on two groups of Jewish isolates, those of the island of Djerba off Tunisia, and those of the Tafilalet Oases between the mountains and the desert in Morocco. As already mentioned, there are two Jewish Haras or ghettos on the island of Djerba. The Cohanim of Hara-Srira have *A* and *B* gene frequencies of 18 and 19 per cent respectively while those of the non-Cohanim of Hara-Kbira are 30 and 15 per cent. This difference is rather similar to that of the two communities of Samaritans in Israel, except that in their case it is the Cohens or Kahins who have the high *A* frequency. Similarly too, the 'White Jews' of Cochin, India, have 40 per cent of *A* genes while the 'Black Jews', associating but not intermarrying with them, have only 5 per cent.

The Jewish communities who lived in the Tafilalet Oases south of the Atlas Mountains in Morocco have long been known but seldom visited. They claim that their ancestors arrived in the time of Nebuchadnezzar. There is no documentary support for this claim, but they are certainly a very old community.

Attention was first drawn by Lévêque (1955) and Mechali *et al.* (1957) to their remarkable blood-group frequencies and especially to their very high frequency of blood-group B. It was thought that this remarkable population merited further examination and therefore, with the cordial cooperation of

Dr. Lévêque, the Cambridge University Expedition to northern Africa visited the region in 1962 and collected blood specimens which were tested at the Blood Group Reference Laboratory in London. At the time of the investigation the Jewish population was spread over an area of 10 000 sq. km, almost entirely stone desert, 100 km south of the Atlas Mountains. They were living in the same villages as the Berbers, but tended to keep apart from them in a separate area in each village. They worked as tailors, as bakers, or at other commercial trades, while the Berbers worked mostly on the land.

The combination of religious and geographical factors had led to a high degree of genetic isolation. To the south is the Sahara desert, to the north the mountains. On the east side is the political boundary with Algeria, and to the west lie 200 km of stone desert before the next river bed is reached. The villages in the region are mainly centred around the dry bed of the river Ziz.

In 1965 Rosenbloom (1966) carried out a historical and demographic study of the same group of Jewish populations. In the short interval since 1962 the situation had changed considerably and he reported that only a few hundred Jews remained in the whole area. This author gives a table of estimates of the Jewish population of the region at various dates from 1920 to 1965, showing populations of many thousands up to 1949 and a catastrophic fall between then and 1965, which he attributes to a declining economy as well as to uncertainty about the future attitudes of Moroccan and Arab nationalist movements. After a certain point the difficulty of living a Jewish life arose in an area with so few remaining Jews. The emigration was first to the larger urban centres of northern Morocco and then to such places as France and Israel. The results of testing the specimens collected by the Cambridge expedition (Ikin *et al.* 1972) confirmed the very high *B* gene frequency, which is one of the highest known for any population, Jewish or other, outside Asia. The gene frequency of 29 per cent found by Ikin *et al.* is now known to be exceeded by the 34·5 and 34·6 per cent of the Karaites of Iraq and Egypt respectively, but there is no historical reason to suspect any close connection between these Karaites and the Tafilalet Jews.

The Tafilalet community has a typical Mediterranean Rh distribution, with the slightly raised *cDe* ($+cD^ue$) frequency of 8 per cent. This is about the same as is found in most other north African Jewish populations, and is evidence of some African admixture, but the level is surprisingly low in view of the position of the oases, just on the edge of Negro Africa. A 4 per cent frequency of the *V* gene, however, suggests a rather greater African component than does the *cDe* frequency.

Their *MNSs* frequencies are also typically Mediterranean, but the total *M* frequency of 69 per cent is higher than is found in other north African Jews. Indeed, their total blood-group picture, apart from the somewhat raised *cDe* and *V* frequencies, would not be out of place in a north Indian population. It is particularly striking that they differ so completely both from the Berbers to the north who have very low frequencies of *A*, *B* and *M*, and from the Negroes to the south, and one can only suppose that the observed frequencies are to a large extent the results of long isolation and genetic drift.

It would be most interesting if members of this former community could be found in Israel and tested for a wider range of polymorphisms.

It has already been mentioned that many Berber communities have traditions and customs which show that their ancestors once practised Judaism. One such community is the Aït Slimane of the Great Atlas Mountains, with an extreme degree of the Berber characteristic of high O (86 per cent) in marked contrast to the Tafilalet Jews.

The Falashas of Ethiopia, despite their Judaism, differ only slightly in a serological sense from their neighbours the Amhara and the Tigré. Their Rh frequencies suggest that they, like other Ethiopian populations, are at least 50 per cent of Negroid ancestry. Since the Amhara have now been examined for a very wide range of serological factors it would be of interest to examine the Falashas more completely for comparative purposes.

No blood samples at all of the Qemant have been tested; they ought certainly to be tested if and when the opportunity arises. The Lemba of Rhodesia, despite their Jewish cultural features give a thoroughly Negroid blood-group picture.

9

THE SEPHARDIC JEWS

THE name 'Sepharad' was used by the prophet Obadiah (v. 20) for one of the lands of captivity of the Israelites, but it was later applied to Spain, so that 'Sephardim' has come to mean 'Spaniards', and the Sephardic Jews may be described briefly as the descendants of that group of Jews who lived in Spain during the Middle Ages but were forced in 1492, unless they accepted Christianity, to leave the country. This was the year of the final victory of the Catholic forces of Ferdinand and Isabella over the Muslim Moors who had for nearly 800 years ruled much of the country. Immediately after the Moors were driven out in battle, the Jews were expelled by decree. The history of the Jews in Spain is, however, much more complicated than is implied by this brief statement. For its elaboration in this chapter I am largely indebted to the work of Roth (1969). The history of the Jews during the period of the Christian reconquest (1300–1492) is dealt with in great detail by Baer (1961).

It is not known when Jews first entered Spain, but the expressed intention of the Apostle Paul (Romans 15:24) to visit that country almost certainly implies the presence there of Jewish communities in the first century of the Christian era. However, long before the Muslin invasion in A.D. 711, the Christianized Visigoths, who had succeeded the Romans in power, had completely suppressed the open practice of the Jewish religion.

Following the initial invasion under Tariq (after whom Gibraltar—Jebel Tariq— is named) the Muslims obtained control of nearly the whole of Spain in the course of four years, and set up an independent caliphate at Cordova.

The Jews who entered Spain in large numbers immediately after the original Muslim invasion came, of course, from north Africa. We have already seen that the Jews of this region, while including many who were of Berber descent, were mostly the descendants of Palestinian Jews who had entered through Egypt. The descendants of these Egyptian Jews had in turn been driven further west by persecution, especially by the Romans, first pagan and then Christian, as long as they controlled north Africa. When the Muslims obtained control they showed much greater tolerance to the Jews of the region, who soon became accepted in the highest official circles. Moreover, the Jews themselves, while retaining their religion, had largely adopted Arabic as their everyday language.

In Spain, where the Muslim rulers had to deal with a Christian native population, as well as with the Latinized rulers and administrators of Christian Europe, the Jews, with their knowledge of languages, were admirably fitted to become administrators. They also excelled as physicians, and as astrologers (who were in great demand).

The freedom of the Jews in Spain and their growing intellectual and religious status soon began to attract Jews from the now downtrodden communities of the eastern Mediterranean and especially, as we have seen (p. 24) from Babylonia which finally, in the eleventh century, yielded the religious and cultural leadership of the Jewish world to Spain.

One of the outstanding figures of the transitional period was Hasdai ibn Shabrut (915–70), court physician to the Caliph Abd-ar Rahman III, who became the latter's chief adviser and negotiator in foreign affairs. His high position enabled him to do much to alleviate the lot of Jews throughout Europe and even in Turkey, and he initiated the correspondence from which we have learned much about the Judaized kingdom of the Khazars between the Black and Caspian Seas (see pp. 45–46). He was also a great scholar; with the help of a monk who converted the original Greek into Latin, he translated the classic botanical treatise of Dioscorides into Arabic and so, through a further Latin stage, made it available to the scholars of western Europe.

About 955 Spain was again invaded by armies from north Africa, this time of Berber origin. The caliphate became broken up into many petty states. The Berbers, as we have seen, had been greatly influenced by Judaism before the Muslim invasion of north Africa and its forced conversions, and their sympathy for the Jewish religion was shown by the high positions which they gave to Jews in their administrations. Again a number of Jews achieved eminence and fame in the combined role of ruler and scholar, notably Samuel ibn Nagrela (993–c. 1056) in Granada and Jekutiel ibn Hassan (assassinated in 1039) in Saragossa, and finally, in the new caliphate which arose in Saragossa, Isaac ibn Albalia (1035–94). Meanwhile, however, Christian principalities were slowly eroding Muslim power and now the Muslims in Spain turned deliberately to north Africa for allies, and called in the armies of the Berber tribes known as the al-Moravides who were fanatical puritan Muslims. They defeated the Christian armies at Sagrajas near Badajoz in 1086. The almoravid rulers set up a new caliphate which at first attempted forced conversion of the Jews of Lucena. Later, Jews once again achieved high office, but continuing military reverses led to a further alliance of the Muslims in Spain with north African forces, this time with the even more fanatical al-Mohades, who insisted on conversion or death for every Jew and Christian, so that the open practice of Judaism was abolished throughout the area which they controlled.

Soon, however, the Christian north began to increase in power, and the reconquest of Spain began. At first the Christian rulers were as intolerant of Judaism as their Visigothic predecessors, but gradually they realized the importance of conciliating the Jews who constituted such an important part of the population of the regions they were invading. Thus once again, for a short time, the Jews gained freedom to live a religious and intellectual life of their own, and went on to produce some of the greatest figures to adorn Spanish Jewry, most notably Moses Maimonides, though as a young man he had to flee to the east with his family before the armies of the Al-Mohades. Not only, however, did Jews achieve high positions at court, but many of them fought bravely and died in the armies of the Christians against the Muslims.

When they came under Christian rule the Jews at first continued to speak Arabic, but gradually they adopted Castilian Spanish which developed into Ladino. Their favoured position continued for about three-quarters of a century. By the early fourteenth century, however, the Muslims were in steady retreat from the peninsula, so that Christian rulers had less need of Jewish help. Thus, largely under papal influence, the tide then turned against them, as it did over most of the rest of Europe.

Repressive laws, and massacres, now began to cause Jews in Spain to profess conversion to Christianity, but large numbers of those who did so, perhaps a majority, became 'Marranos', publicly professing conversion but continuing to practise the Jewish religion in secret. While, however, these 'New Christians' (as converted Jews were called) continued to occupy positions of authority which they had held as Jews, and many even married into 'Old Christian' families, any suspicion that their conversion was not genuine exposed them to the unspeakable cruelties of the heresy-hunting Inquisition, under the Inquisitor-General Tomás de Torquemada, said by some to be of Jewish descent.

Professing Jews, meanwhile, though living a miserably restricted life, could not be called heretics, and so were beyond the power of the Inquisition. Then, in 1492 the Christian armies captured Granada, the last stronghold of the Muslims, all of whom were now expelled from Spain, those who survived mostly going back to their ancestral home, north Africa. In the same year, on 30th March, the Spanish sovereigns, Ferdinand and Isabella, signed a decree expelling all unconverted Jews from Spain within four months. Some 150 000 in number, they had to set out on foot, stripped of nearly all their possessions, for the nearest port or frontier. A great many died on the way. Those who landed on the coasts of Christian Europe found little relief, except for such as reached the papal territories where, rather surprisingly, they were tolerated.

One reason for this was that it suited the Papacy to have under its control a community of Jews who were not bound by Christian rules against usury. As we have seen (and see also p. 16), many of the Roman Jews came direct from Palestine, and considerable numbers had arrived before the Crucifixion of Christ. This enabled the Popes to argue that these, alone among Jews, were not guilty of that crime. Rather illogically, however, the tolerance was extended to all Jews, and not merely in Rome but also in other papal territories, including Avignon at the time of the Great Schism.

Many of the Jewish refugees from Spain reached north Africa where, despite much privation, considerable numbers survived and settled. The most fortunate were those who ultimately attained the Turkish dominions which then included the whole of the Balkan Peninsula, as well as Syria, Lebanon, and Palestine (as they did almost completely within my own memory). In Syria and Lebanon there were still communities of Jews whose ancestors had come directly from Palestine. The few Sephardim who found their way to Safed and Tiberias in Palestine may be regarded as among the forerunners of modern Zionism. The decree of expulsion also applied to the other dominions of the House of Aragon: the island of Sardinia, and that of Sicily where there was an ancient and stable Jewish community.

Those who reached the Portuguese frontier achieved a short respite but, apart from a favoured few who could pay heavily, they were allowed to stay only eight months, and were then forced on board ships and disembarked at the nearest point of the African coast. Those who overstayed

were sold as slaves and many children were sent to colonize the tropical island of S. Thomé where most of them perished. Then the king, Joao, died and his successor Manoel restored liberty to those who had still not been able to leave. Soon afterwards, however, he married Princess Isabella, daughter of Ferdinand and Isabella of Spain, and one of the conditions of the marriage was that all Jews should be expelled. Though many suffered greatly as a result, the main ultimate effect was that of massive forcible baptisms and the creation of large numbers of Marranos; half a century later the Inquisition spread to Portugal and in 1506 two thousand Jews who had professed conversion were put to death as secret Judaizers.

Gradually, from both Spain and Portugal, large numbers of Marranos escaped and, when they reached safe territory, resumed the open practice of Judaism.

To this day, however, there remain, among the Catholics of Spain and Portugal, and in Latin America, communities who, hardly knowing the reason why, continue to perform some of the superficial rites of the Jewish religion (Prinz 1974).

The Sephardim can now very readily be identified by their use of the Castilian Spanish dialect known as Ladino, which they write in Hebrew characters. It is surprising that, having so readily changed from speaking and writing Arabic to Spanish for communication among themselves, they have now for many centuries maintained their Spanish speech against all the pressures of the languages of the countries where they have settled. In Israel they are of course in process of learning Hebrew.

From what has been said about the enforced travels of the refugees from Spain and Portugal it is easy to see why they are now or were recently found mainly in north Africa, Yugoslavia, Bulgaria, Greece, Turkey, Syria, and Lebanon. The older Jewish families of Britain, France, and the Netherlands (formerly ruled by Spain) are also of Sephardic origin.

BLOOD GROUPS

Though there has almost certainly been some intermarriage of Sephardic Jews both with Ashkenazim and with the indigenous inhabitants of the countries where they have settled, yet they show some approach to genetic uniformity, especially with respect to the blood groups other than ABO. This they must have preserved, as they have their language, for the nearly five hundred years that have elapsed since the departure of their ancestors from Spain.

Since many of the Sephardim tested in Israel have been classified only as such, and not as having come from particular countries, these numerous mixed Sephardim provide a useful datum for comparison with other groups of known geographical origin. It must, however, be admitted that their surprising apparent uniformity might prove to be illusory if we could classify these 1900 persons according to their recent countries of origin.

As for almost every other major group of Jews, it is only with respect to the ABO groups that we can classify them at all fully.

In Fig. 4 the *ABO* gene frequencies of the Jews in different countries of south-eastern Europe and the Near East have been plotted. The countries concerned are those which are known to have received large numbers of Sephardic refugees. The total tested Jewish population of each country has been used, whether or not the author of the data has described the subjects as Sephardim. For each country the average gene

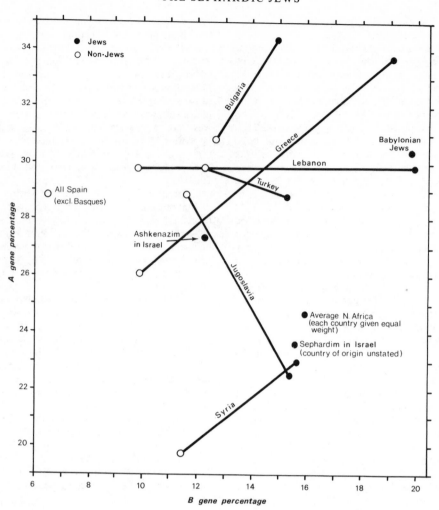

FIG. 4. Frequencies of the *A* and *B* genes in Jewish and indigenous populations of south-eastern Europe and the Near East.

frequencies of the non-Jewish population are also shown, and certain other Jewish gene frequencies have been included for comparison.

In considering the data, and the graphs representing them, it must be borne in mind that, in the confusion of the expulsion from Spain and Portugal, Jews of all origins and classes must have been inextricably mixed. On the small scale groups of Jews from particular localities from Spain have sometimes kept together even to the present day, but each of the receiving countries as a whole is likely to have received a Jewish population of the same average composition, and the relatively small differences between the Jews of these separate countries are likely to have arisen in the main from admixture with the indigenous populations, and perhaps with Jews of other origins. It is unlikely that in the 500 years since the expulsions there has been any large change in gene frequencies due to natural selection.

Thus the point shown as representing the 1959 mixed Sephardic Jews in Israel, with 23·6 per cent of *A* genes and 15·5 per cent of *B* genes, is likely to be very near to the hypothetical one representing the average composition of the original Jews expelled from Spain and Portugal and settling in Africa as well as in Europe and the Near East.

It would be useful to have also a point of reference, how-ever approximate, representing the average composition of the Jews at present (or very recently) living in north Africa (excluding known isolates). The divergence shown in the tables and in Fig. 3 between the averages for the different countries means that this cannot be done with any accuracy. Moreover, both heterogeneity and the widely differing numbers tested do not allow one to treat the whole of them as constituting a single population, and to work out gene frequencies from that. Perhaps the best solution, and the one used here with some reservations, is to take the arithmetic mean of the five national gene fequencies for each of the two genes concerned, and this has been done. The point so obtained falls very near to that representing the 1959 unclassified Sephardim, as does that representing all the Jews in Europe and the Near East specified as Sephardim, but this is perhaps an unbalanced estimate because of the preponderance, among Jews specified as Sephardim, of those from Yugoslavia.

When we come to examine the Jews of the separate European and Near Eastern countries we see that there is a general but rather slight tendency (much less than in north Africa) for the gene frequencies of the Jews to resemble those of the respective indigenous populations, but that, in every case, the Jews have several per cent more *B* genes than the autochthones. Some sets of Jews have more and some less *A* than the non-Jews. Particularly high frequencies of *A* are shown by the Jews of Bulgaria and Greece, both 34 per cent. The *A* frequencies of the Jews in Turkey and Lebanon are also rather high. Since the *A* levels in the indigenous populations of Bulgaria, Lebanon, and Turkey are all rather high, the frequencies in the Jews may be due to mixing with these non-Jewish populations. However, the frequencies found for the Jews of Lebanon are very close to those for the Babylonian Jews, suggesting that the Jewish sample contains a considerable proportion of descendants of Jews who came direct from Palestine (as did those of Babylon). The high *A* frequency found in the Greek Jews may be an accident of sampling due to the small size of the sample tested.

There is, however, a perceptible amount of clustering near the average frequencies already mentioned, and it is most striking that every single national group of Sephardic Jews has a higher *B* gene frequency than the average (11 per cent) for the Ashkenazim.

Even more striking is the difference from the indigenous population of Spain. This country has a very uniform distribution of the ABO blood groups, the average gene frequencies being *A*, 29 per cent, *B*, 6 per cent. The *A* frequency, though outside the range of figures yielded by populations specified as Sephardim, is within the range of national averages for Jews in countries where the Jewish population consists mainly of Sephardim. However, the Spanish *B* frequency is far below that of any known or probable set of Sephardic Jews. A possible partial explanation for the situation is that any tendency towards lowering of the *B* gene frequencies of the Jews during their sojourn in Spain was balanced by the known addition to the community of Jews from Babylonia, with their high *B* frequency (19 per cent in modern Iraqi, i.e. Babylonian, Jews). Despite some discrepancies, it appears that the Sephardim as a whole represent a sample, only slightly modified by recent admixture in most cases, of the original north African Jews. This is particularly the case for their *B* gene frequencies, and we may therefore need to look, for an explanation of the frequencies of this gene, to the two long sojourns of their ancestors among the Egyptians with their high *B* frequency, both before the Exodus, and again at the time of destruction of the second Temple.

The MN groups are less informative than the ABO. The overall *M* gene frequency for all known and probable Sephardic communities is 56·6 per cent, almost the same as in the indigenous peoples of the east Mediterranean countries where they settled, and slightly above the level found in the non-Jewish peoples of north Africa and of Spain. The Jews of north Africa, however, have a somewhat higher average *M* frequency of 59 per cent, and that of the modern Baghdad Jews is 55·9 per cent. All that can be said about the small differences between these averages is that they do not contradict the speculations made above on the basis of the ABO groups.

The Rh blood groups are somewhat more informative than the MN. At the level of D-positive and D-negative, for which we have most information, the frequency of D-negatives is near 10 per cent (32 per cent of *d* genes) for most of the Sephardic populations. The frequency of *d* averaged over a total of 5698 Jews from south and south-east Europe, Turkey, Syria, and Lebanon is 31 per cent; in 714 Jews in Israel, described as Sephardim, without indication of country of origin, it is 34·5 per cent. These figures may be compared with 29 per cent in Egyptian Jews, 33 to 34·5 per cent in other north African Jews (averaged separately over each country), and 28 per cent in Baghdad Jews. Similar frequencies are found in the non-Jewish populations of most of the lands bordering the Mediterranean, including north Africa, southern Italy, the Balkan peninsula, and Turkey. However, the frequency is considerably higher in the non-Jews of Spain, where the average frequency is about 40 per cent.

The Sephardic Jews thus appear to have brought their low frequencies of Rh-negatives from north Africa, if not from Palestine and Babylonia. It is possible, as was suggested for the ABO groups, that a small admixture of genes from the indigenous Spaniards with their high *d* gene frequency, has been offset by the known immigration into Spain of Babylonian Jews who presumably, then as now, had a low frequency of the same gene.

When we consider the results of more elaborate Rh testing, the main feature of interest is the frequency of the *cDe* African marker allele. All Jewish populations which can be classified as Sephardic or probably Sephardic have frequencies of this allele which are well above the 2 to 3 per cent found in most European populations, and which range from 7·7 per cent in those of Bulgaria to 10·7 per cent in those of Yugoslavia and in a mixed Sephardic group in Israel. It will be remembered that similar frequencies were found in north African Jews. Frequencies in indigenous Spaniards average about 3 per cent, but there are a few local concentrations of higher frequency, possibly of Moorish (and perhaps Jewish) origin.

It is, however, certain that the Jews brought these African genes with them into Spain and out again, having acquired them in north Africa or further east. Perhaps Egypt is a likely source, since modern non-Jewish Egyptians have about 19 per cent of the *cDe* allele. It must be observed, moreover, that most of the Sephardic frequencies exceed the 5 per cent level of the Baghdad Jews. Frequencies of *cDE* are all below 14 per cent, showing the normal Mediterranean level as compared with higher frequencies in the more northerly parts of Europe and Asia.

The few data for the other blood-group systems convey little information of value. Frequencies of the *K* gene of the Kell system tend to be higher in the eastern Mediterranean area than in western Europe (where the average frequency is about 4 per cent). The average among the Jews of south-eastern Europe and Turkey is 5·5 per cent.

Frequencies of the Duffy *Fyᵃ* gene maintain European levels. No tests with both anti-Fyᵃ and anti-Fyᵇ appear to have been done on Sephardic Jews, but they do not show the low levels of *Fyᵃ* which would imply the presence of high frequencies of the African marker allele *Fy⁴*.

The few known frequencies of the *Jkᵃ* gene of the Kidd system (52–60 per cent) are mostly near the upper end of the European range. This may be due to the presence of an African component in the Sephardim, since, as already mentioned, frequencies are considerably higher in African Negroes than in Europeans.

The Diᵃ or Diego antigen is characteristic of Asiatic and Amerindian Mongoloids. Few Jews of any community have been tested for it, and no positives have been found anywhere in the Old World, but a Sephardic community in Argentina was found to have 1·5 per cent of the gene. This is presumably the result of intermarriage with American Indians. Though not found in Jews of the Near East, it is present in the Arabs of the Hadhramaut, as a result of movements of these Arabs between Arabia and south-east Asia. One may surmise that it was present in the Chinese Jews.

There are very few data indeed on the frequencies in Jewish populations of plasma protein types, and of most of the red-cell enzyme variants. Such data as exist have been tabulated in this book, but their meaning will remain difficult to assess until more comparative data become available both on other Jewish populations and on their non-Jewish neighbours.

The only enzyme system worth discussing in the present state of our knowledge is that of the variants of glucose-6-phosphate dehydrogenase (G6PD). The gene or genes determining deficiency of this enzyme are almost absent among the Ashkenazim but are present in all other major classes of Jews. G6PD deficiency is present also in most of the indigenous populations of the Mediterranean area. Apart from the Ashkenazim, some of the lowest Jewish frequencies are found in the Sephardim, in whom the average frequency (in males, which is identical with the gene frequency) is about 2 per cent. Unfortunately we do not possess separate data for Sephardic communities in countries in which the deficiency is absent in the indigenous population, and therefore no objective evidence that, like some of their other genetic characteristics, this one was brought with them from Africa into Spain and thence to the countries where they now live.

The frequency of the phenythiocarbamide taster gene in the Sephardim is 49 per cent, virtually identical with that found in Europeans.

From the blood-group evidence taken as a whole, it may be concluded that the modern Sephardim are descended mainly from those Jews who landed in Spain in the eighth century, following in the wake of the Muslim invasions. They presumably drew their genes, in proportions now difficult to ascertain, from the ancient Jewish communities in Palestine, Babylonia, and Egypt, and from the indigenous population of north Africa, including local communities converted to Judaism. A small contribution derived ultimately from Africa south of the Sahara was probably acquired during the sojourn in Egypt after the destruction of the Second Temple. The blood-group data suggest that there was relatively little intermarriage with indigenous Spaniards.

THE JEWS OF ROME

The main Jewish community of Rome, 'The Pope's Jews', is probably descended in part from the Jews who lived there in the centuries before Christianity became the state religion; it has certainly inhabited the present Ghetto for many hundreds of years and has maintained a high degree of endogamy. A blood-group survey was carried out by Dunn and Dunn (1957) but the full results have not been published; they are said to have shown the high frequency of 27 per cent of blood group B.

Though, as the authors could not then fully realize, B tends to be much commoner in Jews than in non-Jews in most countries, this is one of the highest frequencies known among European Jews, apart from the Karaites and Krimchaks of the Soviet Union.

The frequency of the *Cde* allele of the Rh system, 5 per cent, is also among the highest known in Jews.

10

THE ASHKENAZIM

THE term 'Ashkenazim' is derived from the name of Ashkenaz, great-grandson of Noah (Genesis 10:3) and is now applied to the German- (Yiddish-) speaking Jews of northern and central Europe. It referred originally to an area in Anatolia near Mount Ararat, supposed to have been peopled by the descendants of Ashkenaz.

The historical material in this chapter is derived from a variety of sources, but very largely from Roth (1969).

It is generally agreed that the Ashkenazim first developed into a distinct branch of Judaism in mediaeval Poland and Lithuania, but when they first emerge into the full light of history they are already a single people. The circumstances of their early development, and the precise origins of the Yiddish language, have long been matters of dispute.

Most historians have taken the view that the early Ashkenazim of Poland and Lithuania were descended entirely or mainly from the early madiaeval Jews of the Rhineland, but there have long been a few who have regarded part or the whole of their ancestry as being derived from lands to the south-east, either from the Jews, presumably mainly of Palestinian origin, who lived in such places as Persia and Bokhara, or from the Jewish proselytes of Turkish ancestry of the Khazar empire.

When, some twenty years ago, I first tried to analyse the blood-group data then available (Mourant 1959) I was already partially aware of these diverging views, and I thought that I could detect evidence for an eastern derivation of some Jewish communities.

When, more recently, I attempted a graphic analysis of the much more extensive data tabulated in this book, the work of Koestler (1976) on the origin of the Ashkenazim had just appeared, and I then regarded it as more fully authoritative than I now should. I think, however, that it may add appreciably to the weight to be attached to the results of my own analysis if I explain the circumstances under which it was carried out: it was planned largely for the specific purpose of testing Koestler's theory of a Khazar ancestry for the early Ashkenazim, but I was forced by the blood-group data alone to the conclusion that the Ashkenazim were essentially European Jews closely related to the Sephardim. It is for this reason that, in revising this chapter, I have retained a large part of what I first wrote about Koestler's theory, though I have since been introduced by Professor C. Abramsky to much of the detailed evidence for a western origin, of which I was previously unaware.

The main facts have been well summed up in a long review by Abramsky (1974), published, it should be noticed, before the appearance of Koestler's book, though he (Abramsky 1976) has also, rather more briefly, reviewed Koestler's book itself.

In the early centuries of the Christian Era there were Jewish communities in all the lands immediately north of the Mediterranean, and in the Persian dominions. The problem of the origin of the Ashkenazim is largely that of finding out when and where the Jews, either as colonists or as proselytizers, crossed the line of mountains and seas separating these lands from the plains of central Europe and central Asia.

It is said that, by the third century A.D., there were already Jews in France, Dalmatia, Scythia, and the Crimea, and considerable numbers in Germany, mainly in the Rhineland, large enough for them to have been the subject of special legislation.

For several hundred years, during the so-called Dark Ages, we then almost lose sight of the Jews in these countries, though the Sephardim were flourishing in Spain; it is likely that so long as the power of imperial Rome persisted, and possibly even later, the Jewish colonies in Italy were supplying migrants into France and Germany.

Further east we know that the Jews of Persia were crossing the mountains and seas into what is now Soviet central Asia.

One very important event was the conversion of the rulers, and probably of many of the ruled, of the kingdom of the Khazars of the Caspian region to Judaism. As already stated, the following account was drafted before I was aware of much of the still scanty information on the early Jewish settlers in Poland and Lithuania. I have retained most of the original draft, adding a few more facts and correcting others, and omitting some details now seen to be irrelevant.

THE KHAZARS

The Khazars were a large tribe, or perhaps the dominant member of a group of tribes. They are said to have spoken a language of the Turkish group, related to modern Chuvash. They appear to have moved westwards from central Asia in the fifth century A.D., to occupy an area between the Black and Caspian Seas and bounded on the south by the Caucasus mountains, an area which they dominated by the second half of the sixth century.

The Muslims, originating in Arabia, had spread northwards and north-eastwards through Mesopotamia and Persia during the eighth century, until the Khazars blocked their further advances into the Don and Volga basins whence they might otherwise have entered the Balkan countries, outflanking the Christians of Byzantium.

For several centuries the Khazars also blocked the southward advance of the Russians, and split the Bulgar peoples, one branch of whom remained on the Volga while the other was driven westwards into the Danube basin.

About A.D. 740 the king, or Kagan, of the Khazars formally adopted the Jewish religion which then spread at least to all the ruling classes.

The story of the conversion of King Bulan was written at some length by one of his successors, King Joseph, more than two hundred years after the event, in a letter to the Spanish Jewish statesman Hasdai ibn Shaprut. From it we learn that Bulan adopted Judaism only after listening to the arguments of exponents of the three great monotheistic religions. Joseph makes no claim for descent from Abraham but says that the Khazars are descended from Japhet, son of Noah (and not, like the Jews, from his brother Shem). If

Bulan was determined for political reasons to adopt mono-theism, then Judaism had the advantage that he would not in any way be subject to the ruler of either of the politically powerful theocracies, the Christian Emperor at Byzantium or the Muslim Khalif of Baghdad. By contrast the head of the eastern Jews, the Exilarch or Resh Galutha in Mesopotamia, had little political power.

It is of some importance to know whether the Khazar Jews were virtually all converts (or descendants of converts), or whether any considerable numbers of members of some older body of Jews had been absorbed into the Khazar community. We know that, as just mentioned, religious leaders went by invitation to instruct them in the practices of Judaism, but these may have been few in number. The admission of their own King, that they were descendants of Japhet, not of Shem, suggests that no substantial numbers of Semitic Jews had entered the community.

A further point of interest is that the oldest copy of Joseph's letter is in the Firkowich collection of Hebrew documents in the Leningrad Public Library. Most of the documents in this collection are known to have come from the famous Cairo Geniza, which was part of a Karaite synagogue. According to Koestler, Firkowich is known to have doctored some of his documents in order to support his case that the Karaites differed racially from orthodox Jews and that, unlike Jews by race, they ought not to suffer adverse discrimination at the hands of non-Jews. The Khazar document appears, however, not to have been altered, and its probable origin gives some support to the view that the Khazars had Karaite connections.

In the middle of the ninth century A.D. the Khazars built the fortress city of Sarkel on the lower Don, to protect themselves from the growing power of the Rus or Russians who were extending their domains southward. The Rus are generally regarded as of Viking origin, and their highways were the great rivers of Russia, leading to the Caspian and Black Seas. The capital city of the Khazars was at Itil at the head of the delta of the Volga. To the west, Kiev on the river Dniepr was for a time within the region subject to the Khazars, but was taken by the Rus about 862.

There followed several centuries of considerable population movement in eastern Europe, marked by the great expansion of the Russian dominions and the gradual loss of power by the Khazars. At the same time the Magyars and the western Bulgars, formerly in the Khazar zone of influence, moved into the regions which they still inhabit, though the eastern Bulgars remained on the Volga. In 965 the Khazar capital, Itil, was destroyed, apparently by the Russians, but was probably rebuilt.

The following centuries were marked by great confusion. The only single more or less consistent source of historical information is the Russian Chronicle, written by monks and inevitably from the Russian and Byzantine Christian point of view. Koestler has assiduously collected every scrap of information having any possible bearing on the history of the Khazars. For details and bibliographic references the reader must consult his book. The general impression conveyed by it is that the Khazars, though suffering still further defeats by the Russians, remained a distinct nation and retained the Jewish religion for several hundred years. Indeed, as late as the 'Khazar correspondence', about A.D. 960, the Khazar King writes: 'I guard the mouth of the river and do not permit the Rus who come in their ships to invade the land of the Arabs. . . . I fight heavy wars with them.' One important event for the Khazars was the conversion of

Prince Vladimir of Russia to Christianity in A.D. 989. This is said to have been preceded, like the conversion of the Khazar King to Judaism, by the visit of Muslim, Jewish, and Christian delegations, but this time separate ones from the Roman and Byzantine churches.

Now Byzantium, which had formerly sided with the Khazars against the Russians, became the ally of Russia but, for the time being, was unable to help her greatly in the face of a new enemy. The weakening of the Khazar kingdom allowed the Polovtsi, or Kumans, an offshoot of the Turkish Ghuzz who lived between the Aral and Caspian Seas, to sweep across the steppes, between the Russian and Khazar homelands. They finally reached as far as Hungary, and dominated the area from the late eleventh to the thirteenth century. Meanwhile, another branch of the Ghuzz, the Seljuk Turks, began to take over Anatolia from Byzantium.

The Polovtsi were followed by the Mongols of the Golden Horde who also penetrated as far as Hungary. With their advent and the resulting confusion, we almost completely lose sight of the Khazars, but the few Russian mentions of them still refer to them as Jews, while Muslim sources suggest that many of them had turned to Islam.

In 1245, the Mongol invaders had established their capital at Sarai Batu on the Volga Delta, on the site of the former Khazar capital of Itil. Yet the papal envoy, Joannes de Plano Carpini, who mentions this, also refers, among the people of the northern Caucasus, to the 'Khazars, observing the Jewish religion'.

The Mongols carried out extensive massacres and destroyed the irrigation network upon which Khazar farming had depended. Not only was Khazaria devastated, but also the southern parts of the Russian domain, including Kiev, and the lands of the eastern Bulgars. Then followed the Black Death of 1347–8, which killed a large number of the survivors and increased the devastation of these lands, a devastation which has only recently been remedied.

We shall have to consider what happened to the surviving Khazars and their Jewish culture, but first we must return to the west and try to envisage what was happening to the Jews in central and western Europe.

WESTERN AND CENTRAL EUROPE

The accession of Charlemagne (born 742, reigned 768–814) who ruled a large part of France, Germany, and northern Italy, may be regarded as marking the end, in western Europe, of the confusion of the Dark Ages and the beginning of the Medieval period. He encouraged the settlement of Jews both in France and the Rhineland.

Up to the time of the Crusades the French and Rhineland communities flourished and probably expanded, despite minor episodes of persecution. There was continuing contact in Provence between the Jews of Spain and of France. At the same time Jews were pressing into more southerly and easterly parts of Germany, apparently from the south, perhaps over the Alpine passes from the colonies in northern Italy and Dalmatia. Under the Norman kings Jews also entered England.

In 1095 Pope Urban II summoned Christendom to recover the Christian holy places of Palestine from the hands of the Muslims, and so initiated the series of wars, lasting many generations, known as the Crusades. But at the same time and in close association with the crusading urge, the tradition spread, often with the official support of the Church, that the Jews as a race were guilty of murdering Christ, and that it was therefore meritorious to kill them. There followed massacres in the Crusaders' wake across Europe and the Near East, and in the Holy Land itself. Such massacres continued to take place throughout the whole period of the Crusades, and indeed for many years afterwards.

One particularly heavy blow to the Jews of Europe was the Black Death of 1347-8. Not only did it almost certainly kill about half the community, but its transmission was widely attributed to the malevolence of the Jews, and made an excuse for still further bloodshed.

While England and France were each already in the thirteenth century more or less united monarchies, so that the expulsion of the Jews was almost completely effective, Germany was still a collection of small principalities and, despite persecutions and massacres, there was usually some place not far away, to which expelled Jews, and survivors of massacres, could go. Yet it might have been better for them had they been expelled completely from the German region for, like cultivators returning repeatedly to their devastated vineyards on the side of a volcano, they were condemned to a life punctuated by vast exterminatory killings in one town after another. Then, for the Ashkenazim, the Renaissance and the Reformation brought a short respite, just as the Inquisition was beginning to torture the Sephardim.

Meanwhile, in the middle of the thirteenth century, the rulers of Poland, devastated after the final retreat of the Mongol invaders, were trying to restore order and prosperity by encouraging Germans to settle in the cities; with them, almost certainly, came Jews. It is however impossible to deduce, from the scanty historical records, how many Jews may have migrated from Germany to Poland, and hence what proportion of the members of the various Jewish communities found in Poland and Lithuania in the fourteenth century were the descendants of Jews who had entered from Germany and other western countries.

THE JEWS OF POLAND AND LITHUANIA

There is no doubt that the main ancestors of the modern Ashkenazim were the Jews who, in considerable numbers, lived in Poland and Lithuania in the late Middle Ages, and who spoke Yiddish, a dialect which consisted mainly of German words, together with some Slavonic and Hebrew ones, and which was written in Hebrew characters. The precise nature and origin of this language is of the utmost importance in any discussion as to the origin of the Ashkenazim themselves, and this is further considered below.

Most historians have held the view that the German-speaking Jews of Poland and Lithuania were descended from the earlier communities in western Germany. Koestler (1976), on the other hand, claims that they are descended almost entirely from the Turkish Khazar proselytes. He claims no originality for this view for which he quotes many previous authorities but, as already mentioned, he had made what is certainly the most intensive effort yet attempted to bring together all the relevant evidence. I am not a professional historian, nor have I been able to consult many of the primary sources mentioned; here I cannot do more than try to summarize the main historical arguments, as an introduction to the discussion of the genetic evidence.

On the one hand Koestler claims that the initially small and ultimately decimated German Jewish communities of the early Middle Ages could not possibly have produced the numbers of emigrants needed to give rise to the large Polish-Lithuanian community. Moreover, he claims that the Yiddish language, further discussed below, is derived not from a western but from a south-eastern form of German.

On the other hand the once flourishing Khazar nation, of central Asiatic origin but Jewish by religion, disappears from history in the Middle Ages, following the Russian and then the Mongol invasions, and the Black Death. Soon after this, large numbers of Jews, whose ancestry is not precisely specified, appear in Poland. Abramsky considers that the numbers were by no means as large as Koestler claims, and that they can readily be explained by migration from the west. There is, according to Koestler, evidence of a substantial migration of Khazars, who were presumably Jewish by religion, to Hungary about A.D. 900, but virtually nothing about any such incursion into Poland.

The Polish (and Lithuanian) Jews did however develop a culture and mode of life somewhat different from other groups of Jews, based on the *shtetl* or small town, rather than the ghetto, or shut-in Jewish town quarter familiar elsewhere, and their synagogues had a pagoda-like shape. They were also great coach-builders and masters of coach transport.

At about the same time as the arrival or great expansion of the Rabbinic Jewish community of Poland, large numbers of Karaites arrived in Poland, who were the ancestors of the modern Karaites of Lithuania and the Ukraine; there is clear evidence that they were deported members of the Karaite community of the Caucasus. They spoke, and their descendants still speak, a Turkish language. Koestler quotes the Russian census of 1897, according to which there were 12 894 Karaites living in the Tsarist empire (which then included Poland, Lithuania, Ukraine, and Crimea). Of these 9666 spoke Turkish (equated by Koestler with the Khazar language), 2632 spoke Russian, and only 383 Yiddish.

Linguistic evidence, together with our knowledge that some at least of the Khazars north of the Crimea followed Karaite practices (as observed by Rabbi Petachia about 1180) suggests that the present Karaites of the Crimea and Lithuania are of Khazar ancestry.

For the Rabbinic Jews of Poland on the other hand we have neither historical nor linguistic evidence of a Khazar origin.

THE YIDDISH LANGUAGE

The main unifying feature of the Ashkenazim, apart from the Jewish religion itself, is the Yiddish language. Koestler stresses the east German origins of this dialect, which in fact contains elements from western Germany as well as from the Bavarian area—it even contains some French words. It contains no Turkish words and it is easy to account for the Slavonic words without assuming eastern origins. The early Slavonic words are all from the Czech language, and are derived from rabbinical 'responsa', verbal statements made by non-Jews, in their own language and taken down phonetically in Hebrew characters, at a time when, in fact, the Czech language had not yet been written down in the Roman alphabet. Polish words do not appear until much later and are certainly not an original feature.

With regard to the Bavarian element, it should be noted that there were, in the early Middle Ages, many Jews in northern Italy and the Balkan peninsula, and indeed, as we have seen, at many other places along the great Eurasian mountain backbone from Byzantium to the Caucasus and Persia. In the Austrian provinces of Carinthia and Styria we

48 THE ASHKENAZIM

find numerous places called Judendorf, Judenstadt, Judenburg. By the end of the fifteenth century Jews were expelled from both provinces and went to Italy, Hungary, and Poland. But perhaps even before that they were crossing the Alps, into Bavaria, and acquiring the local variety of German. Then, as the kings of Poland encouraged the settlement of Germans, it is likely that many Jews came in with them. It is possible that there was already a substantial Jewish community in Poland, of more easterly origins, which would have welcomed the teachers of Rabbinic Judaism and Jewish literary culture included among the newcomers.

However, even if many of the Jews in Poland had come from the Black Sea–Caucasus–Caspian region they were not necessarily all Khazars. We have seen that Jews, ultimately almost certainly of Palestinian origin, had penetrated much of south-west Asia and crossed the barriers of the Caucasus and the mountains of Kurdistan and Persia into the steppe region. Perhaps, indeed, they may have mixed in some numbers with the proselyte Jewish community of Khazaria. The question of the origin of the Ashkenazim of Poland is thus a most complex one, and, as we shall see, it is one to the solution of which the blood-group evidence that we now possess makes a substantial though not completely unambiguous contribution.

Whatever may have been its origin, the Jewish community in Poland flourished from the fourteenth to the seventeenth century and suffered little from persecution, so that this country was for many years the main religious and cultural centre of world Judaism.

The turning point came during the prolonged wars between Poland and Russia, which lasted for much of the seventeenth century. At the beginning of the century the joint kingdom of Poland and Lithuania ruled the Ukraine, and continued to do so until, in 1653, the Cossacks of that region rose against their Polish masters, massacring both Poles and Jews. The Cossacks appealed to Russia for protection and, as a result, the Ukraine east of the River Dnieper became incorporated in Russia. From this time onwards, as the Russians advanced into Poland, more and more Jews came under Russian rule and suffered systematic repression, punctuated by pogroms. Finally, about the end of the eighteenth century, Poland, with its large Jewish population, was completely divided up between the empires of Russia, Germany, and Austria-Hungary.

Now began the mass movement westward of the Ashkenazim, first into Germany, a movement which after long tolerance and even prosperity, led to the greatest tragedy of all, and then to freedom in Israel, western Europe, and America. Demographically speaking, these movements must have caused considerable mixing, within the Ashkenazim as a whole, of Jews of various geographical origins. Certainly, however, the mixing has not been complete, and one of the objects of blood-group analysis has been to distinguish persisting geographical strains among the present communities.

THE ABO BLOOD GROUPS

Once again the main discussion of blood-group frequencies must depend upon the ABO groups. Though the historical movements in the last hundred years have been mainly from Russia to central and western Europe, it will be convenient here to consider first the situation in countries outside the Soviet Union, before embarking on the greater complexity of the situation within the U.S.S.R.

Figure 5 shows the A and B gene frequencies of Jews in those countries of central and western Europe where the Jews are known to be mainly Ashkenazim. Unfortunately we have only very scanty data for western Europe, and none at all for several countries. In each country for which we have data the frequencies found in the Jews are compared with those of a representative sample of the non-Jewish population.

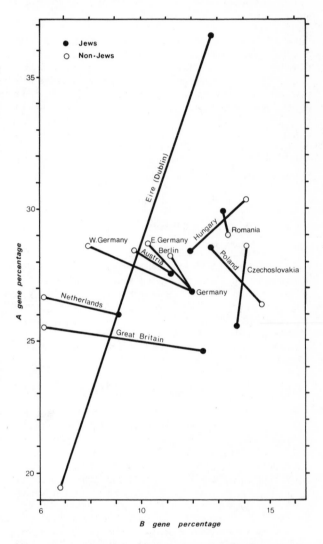

Fig. 5. Frequencies of the A and B genes in Jewish and indigenous populations of northern and central Europe.

It will be seen that there is a closer similarity between the Jewish populations of the various countries than between their non-Jewish inhabitants. The Jewish gene frequencies cluster around the values A, 27 per cent; B, 12 per cent. The average frequencies for 5573 Ashkenazim in Israel, whose countries of origin are not stated, are very near these values, with A, 27·4 per cent; B, 12·4 per cent.

There is a slight tendency for the differences between the Jewish communities from the various countries to run parallel to those between the non-Jewish inhabitants, but there is not a relatively uniform tendency, and as is found in north Africa and among the Sephardim of south-east Europe and the Near East, for the Jews to have higher B frequencies than

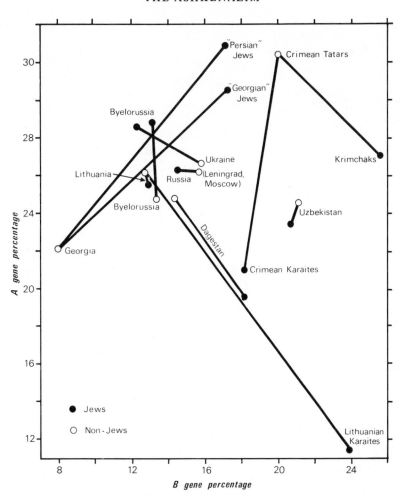

FIG. 6. Frequencies of the *A* and *B* genes in Jewish and indigenous populations of the Soviet Union.

the non-Jews. The Ashkenazim in the more westerly countries have higher *B* but lower *A* frequencies than the non-Jews, whereas further east the non-Jews have the higher *B* levels, and the *A* frequencies may be higher or lower. The marked anomaly in the case of Ireland is clearly due to sampling error in the very small sample of Jews tested.

The ABO data suggest that a relatively uniform Jewish population emerged from the lands that now constitute the Soviet Union, and became only slightly modified by mixing with the populations of the new host countries.

In the Soviet Union itself the picture at first sight appears much more confused, with very little resemblance between gene frequencies in Jews and non-Jews (Fig. 6). There can be no doubt that this is due to considerable variation in the origins of the Jewish communities which inhabit the different parts of the Union, even though we do not in all cases know what these origins were.

If we consider first the main European area, consisting of Great Russia, Byelorussia, Ukraine, and Lithuania, where most of the Jews are Ashkenazim, there is a clustering of Jewish gene frequencies very near to that found further west; the mean *A* frequency is again about 27 per cent, but that of *B* is slightly higher than in the west, near 14 per cent. These values are very near those of the non-Jewish populations, but there is little sign of any correlation between the frequencies

shown by Jews and by non-Jews of each region within the area as a whole.

All the other Jewish communities in the Union have higher *B* frequencies than those just mentioned, but both their *A* and their *B* frequencies have a very wide scatter.

In Georgia both the 'Georgian' and the 'Persian' Jews have very much higher *A* and *B* frequencies than the Georgians themselves. The Jews of Dagestan, thought to have come from Persia, have considerably more *B* but less *A* than the indigenous population, while those of Uzbekistan, also thought to be from Persia, closely resemble the indigenous population.

The Karaites of Lithuania have extremely high *B* and low *A* frequencies, utterly different from those of the orthodox Jews of the same region, which closely resemble those of the indigenous Lithuanians.

In the Crimea there are wide differences between the non-Jewish Tartars, the orthodox Jewish Krimchaks and the Karaites, though all have rather high *B* frequencies. The Crimean Karaites, while on the whole resembling those of Lithuania, have less extremely low *A* and high *B*. In this case however we have fairly good historical evidence of their relationship to one another. Perhaps one or both communities have been affected by genetic drift.

In view of the suggestion that both the Karaites and the

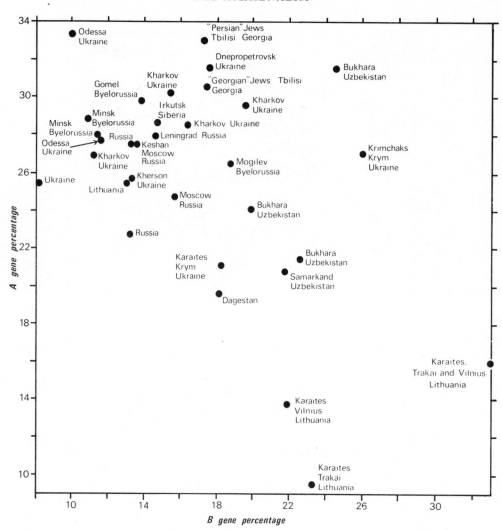

FIG. 7. Frequencies of the *A* and *B* genes in individual Jewish communities of the Soviet Union.

Ashkenazim are descended from the Khazars (a problem further discussed later), and of the very wide differences found between the blood-group frequencies of different Jewish populations in the Soviet Union as a whole, their *A* and *B* gene frequencies were plotted in a new graph, giving a separate plotting to the results of each published survey (Fig. 7). The object was to see whether this device would in any way bridge the gap between the two main sets of Jews. The effect was to open up the cluster of Ashkenazim results, though only to a slight extent, but a resemblance now became clearer than before between the Jews of Dagestan and Uzbekistan and the Crimean Karaites.

This is about as far as we can go in considering the ABO groups of the Ashkenazim and other Soviet Jews by themselves. Before we extend this part of the discussion to a wider area we must look at the other blood-group systems as they affect the Ashkenazim.

THE MN GROUPS

In the countries where the majority of the Jews are or were Ashkenazim, the *M* gene frequencies range from 54·5 to 61·1 per cent, levels mostly slightly above those of the

indigenous populations, and showing a slight tendency, parallel to that of the non-Jews, to increase from west to east. A substantial sample of mixed Ashkenazim in Israel shows only 53·5 per cent of *M* genes but this is probably because they came mainly from Germany and Poland where the average *M* frequencies for nationally identified Jewish populations are 54·5 and 55·4 per cent respectively. The average *M* frequencies in the Ashkenazim thus do not differ greatly from those found for Sephardim or for north African Jews. The *M* frequency in the Karaites of Lithuania, on a small sample, is 69·6 per cent, which is considerably higher than that for any known Ashkenazi population.

Frequencies of the *S* gene, in Ashkenazi populations, where known, are within the normal range for non-Jews of the countries concerned, but the few available data do not show any marked regional trend.

THE Rh SYSTEM

The Rh groups are more informative. The frequency of the *d* gene in Jews of the European countries concerned varies from 25·3 to 31·7 per cent, and similar figures, within or near this range, are found for groups of presumed Ashkenazim in the New World and in Israel. The average d gene frequency for all tested Jewish populations which are known to be, or are probably, Ashkenazim, is 30·1 per cent. This is somewhat lower than for Sephardim or for most north African Jewish populations and is, of course, very much lower than for the indigenous populations of any of the European countries concerned (which have higher frequencies of d than Mediterranean European countries). Again the Karaites of Lithuania are the exception, with 37·0 per cent of d on a small sample.

When we examine the results of more elaborate Rh testing, we find that there is still a considerable degree of resemblance between the various Ashkenazi populations tested. Frequencies of CDe (R^1) are mostly about 50 per cent, systematically slightly higher than in the Sephardim and considerably above the levels found in non-Jews of central and northern Europe, while that of cDE (R_2) is about 12 per cent, slightly higher than in the Sephardim but somewhat below frequencies in the host populations. The African marker allele cDE (R_0) is always present, and at frequencies markedly higher than in indigenous European populations, the average being about 7 per cent. This is, however, somewhat less than in the Sephardim.

OTHER BLOOD-GROUP SYSTEMS

The Kell gene K has frequencies between 6 and 8·3 per cent in most Ashkenazi populations; these levels are distinctly higher than are found in most European populations, and are also somewhat higher than in the few Sephardi populations tested. Similar relatively high frequencies are found in the Arabs of Arabia (and in Iranian Jews, but not in the Yemenite Jews from Arabia).

Frequencies of the Duffy gene Fy^a are mostly near 43 per cent, which is near the average for Europe. Only one small sample, of German Jews, has been tested both with anti-Fy^a and anti-Fy^b. No homozygous double negatives were found but the deficiency of apparent heterozygotes could indicate the presence of a small percentage of the African marker Fy^4 gene.

The Kidd Jk^a gene has frequencies which are mostly near 60 per cent, which is above European levels. As this gene has still higher frequencies in Negroes, this may be another indicator of an African component in the Ashkenazim.

The very rare Radin (Rd) antigen has so far been found only in Scandinavians and Jews, a rather surprising combination. In each case the frequency has been below 1 per cent. The Jewish populations tested were mixed ones, probably mainly Ashkenazim. It is desirable that tests for this antigen should henceforth be carried out whenever Jewish blood samples are being examined, as the Rd gene may prove to be a useful marker for populations of particular origins.

PLASMA PROTEINS AND RED-CELL ENZYMES

Some Ashkenazim have been tested for a wide range of plasma proteins and red-cell antigens, but in most cases the numbers tested are so small, and there are so few data available for comparison, that no discussion is called for.

For the haptoglobin system Ashkenazim in Israel have 31 per cent, and Polish Jews, on a small sample, 38 per cent of Hp^1 genes, both of which figures are lower than the general European level of about 40 per cent. The average for non-Jews in Poland is however 37 per cent, and levels are in general lower among non-Jews in the near East and south-eastern Europe than in the rest of Europe.

Considerable numbers of Ashkenazim as well as of other Jews have been tested for hereditary deficiency of the red-cell enzyme glucose-6-phosphate dehydrogenase (G6PD). The Ashkenazim are the only large Jewish group to have less than 1 per cent of the gene for this deficiency; their gene frequency is 0·37 per cent.

Since this otherwise rather harmful gene gives protection against malaria, it may be that the Ashkenazim, who have long lived in less malarious areas than most other Jewish groups, did have a higher frequency at one time but have gradually shed the gene by natural selection.

THE ASHKENAZIM AND THE SEPHARDIM

We have already seen that the Sephardim and their ancestors have a continuous history extending back some 2000 years, and their blood-group picture is consistent with their having originated in Palestine, sojourned in Babylonia, Egypt, north Africa, and Spain before dispersing to various countries of south-eastern Europe and the near East, and having, probably via Egypt, picked up a small percentage of African Negroid genes. The one feature that is somewhat difficult to explain is their rather high frequency of the B gene. Perhaps this comes from their having mixed with the Babylonians or the Egyptians, or both.

Looking at the complete blood-group picture of either the Ashkenazim or the Sephardim separately, one may observe that neither of these populations resembles closely the peoples among whom they now or recently have lived, and the range of variation between separate samples of the Ashkenazim compared with one another or the Sephardim compared with one another, is so small that we can be sure that each is essentially a single population group.

When, however, we compare Ashkenazim with Sephardim we find that there are indeed systematic differences between them. But these are so small that we can hardly avoid the conclusion that the two populations have a common origin, and a common original blood-group picture, only slightly modified in one direction or another by their different histories since separation.

A close resemblance with respect to one blood-group system might be accidental, but there is a resemblance extending to all the blood-group systems for which data are adequate. They are not identical, but frequencies for each system differ by just a few per cent. This, taken together with their common traditions, can hardly mean anything other than a common origin. To prove that the Ashkenazim had an origin independent of that of the Sephardim would need the demonstration that the former had at least as close a resemblance with some other and non-Palestinian population group as they have with the Sephardim. To establish the Khazar hypothesis in particular, such a resemblance would need to be shown with a modern population demonstrably closely related to the ancient Khazars.

Before, however, we further examine Koestler's Khazar hypothesis, we must look at the systematic small differences between the Ashkenazim and the Sephardim, and see whether we can account for them in any other way than that proposed by Koestler.

The Ashkenazim have very consistently several per cent less B genes than the Sephardim, and on the whole, but less consistently, several per cent more A. The difference with respect to B is likely to be due in part to extra B genes acquired by the Sephardim, probably from Egypt, since they separated genetically from the Ashkenazim. It is probably also partly due to the relatively low *B* frequencies present in the populations among whom the Ashkenazim have lived (though it must be admitted that the Sephardim too have for a long time lived among peoples with similarly low *B* frequencies). The rather high *A* of the Ashkenazim may, on the other hand, have come from the peoples who live along the main east–west mountain chain of Europe and Asia, which they must have crossed to reach the plains to the north. If the Ashkenazim had received many of their ABO genes from a central Asiatic source one would have expected much higher *B* frequencies.

The slightly higher *M* frequency in the Ashkenazim than in the Sephardim can only be explained by guesswork. The indigenous *M* gene frequency is indeed higher in eastern Europe than in north Africa but why then do the Jews of north Africa have a somewhat higher *M* frequency than even the Ashkenazim?

For the Rh system the rather low *d* frequencies of both Ashkenazim and Sephardim are similar to values found in the indigenous peoples of the eastern Mediterranean area; but why then are they lower in the Ashkenazim living among people with a high *d* level than in the Sephardim who have lived among peoples with a lower level (except in Spain, where the ABO data suggest that there was virtually no mixing with the Spaniards)? A low *d* might have come from central Asia but it would probably have brought a high *B* with it.

The slightly higher level of *CDe* in the Ashkenazim than in the Sephardim is difficult to explain, as frequencies of this allele are higher in the Mediterranean area than in northern and central Europe; the higher *cDE* also found in the Ashkenazim accords with the northward increase of this allele in the indigenous populations.

We have, however, so far been discussing minor variations in frequencies of genes which do not differ greatly in their frequencies throughout Europe. Much more significant is the African marker allele *cDe* which, as might be expected, reaches higher levels in the Sephardim than the Ashkenazim. Though the *cDe* levels are, in general, higher in indigenous populations in Asia than in Europe, they never approach the African levels, and it is difficult to see how either set of Jews could have acquired its present level of this allele except directly or indirectly from Africa while living somewhere in the Mediterranean area.

The evidence for considerable hybridization between Jewish populations and their slaves, largely Africans, has recently been considered in some detail by Patai and Wing (1975) who have made it much easier than it was, in terms of social history, to account for the presence of the various African marker genes.

THE NON-ASHKENAZI JEWS
OF THE U.S.S.R.

Those Jewish communities in the Soviet Union which do not speak Yiddish have, in most cases, as we have seen, ABO frequencies which differ markedly from those of the Ashkenazim, and we must now explore the possibility of any of these being the descendants of the Khazars.

Unfortunately we have little indication of what may have been the blood-group frequencies of the Khazars. If they indeed came from central Asia then they are likely to have had, like the present inhabitants of that region, high fre-

quencies of *B* and *M* and a low frequency of *d*. They were, however, Turkish speakers, and the modern Turks of Anatolia have a high frequency of *A*, rather low *B*, moderate *M*, and high *d*, not unlike average Europeans. But these modern Turks are a mixture probably incorporating a high proportion of genes derived from the pre-Turkish population who were largely Armenians whose ancestors had entered Anatolia from Europe.

The Turkish speakers of Iran, and the Turkomans of the Soviet Union, however, do show high frequencies of *B* genes, averaging about 20 per cent, with 20 per cent of *A* in Iran and 27 per cent in the Soviet Union. Those of Iran have approximately 52 per cent of *CDe*, 16 per cent of *cDE*, 5 per cent of *cDe*, and the relatively low figure of 25 per cent of *d*.

It must, however, be remembered that one early chronicler describes the Khazars as pale-skinned, blue-eyed, and often red-haired (Ibn Said al-Maghribi, cited by Koestler 1976), which, if true, might imply a European blood-group picture.

The Jews of Georgia, both Georgian and Persian speaking, must have entered the present Soviet Union from the south and they have, in any case, extremely high frequencies of *A* and rather low *B*, thus perhaps relating them to the Armenians and the presumed hybrid Anatolian Turks. The Jews of Bukhara, Samarkand, and Dagestan, with ABO frequencies not too remote from those of the Turkomans, almost certainly entered from Iran, though it is possible that they may have intermarried with Jewish Khazars. Thus the main Jewish populations for which we have to consider the possibility of a predominantly Khazar ancestry are the Karaites of the Crimea and Lithuania, and the Krimchaks or orthodox Jews of the Crimea.

All the Karaite populations tested have high *B* frequencies. Their *A* and *B* frequencies show considerable scatter, due no doubt partly to genetic drift and partly to sampling error, but the frequencies found are certainly near enough to those of the Turkomans, and to central Asiatic frequencies in general, to make a Khazar ancestry definitely possible. The orthodox Krimchaks, with high frequencies of both *A* and *B*, lie right outside the Ashkenazi cluster and would fit reasonably well, considering the smallness of the sample, into the central Asiatic picture.

It must be admitted that the Rh frequencies of the Iranian Turks show a certain resemblance to those of the Ashkenazim, but these Turks, especially in their very low *d*, have a distinctly more eastern aspect.

THE ORIGIN OF THE ASHKENAZIM

The general conclusion to be drawn from the blood-group data is that the Ashkenazim are essentially a single population, largely, if not mainly, of Palestinian Jewish descent. The strongest evidence is their systematic resemblance to the Sephardim, whose Palestinian ancestry is more firmly established by historical records. As already stated, these two major groups of Jews differ distinctly with respect to each main blood-group system, but the difference for any one system is never very great. This is what one would expect for two populations of common origin but separated for not much more than a thousand years. The incorporation of a Khazar component in the Ashkenazim cannot, however, be ruled out completely.

On the other hand, their blood-groups show a strong possibility that all the Karaite populations of the Soviet Union, as well as the Krimchaks, may be largely of Khazar ancestry.

11

GENETIC ASPECTS OF DISEASES IN JEWS

Diseases which strike men originate from three causes: first, heredity from father to son, as in gout, according to natural scientists; secondly, from a place or climate, water or air, which produces goiter in men, as seen in Vera di Placentia and the city of Buitrago; thirdly, through close and perpetual contact, in communicable diseases, such as bubonic plague and other ailments.

Toledat Yitzhak, commentary on the Pentateuch, by Rabbi Isaac Caro, expelled from Spain in 1492. (Translation from *Israel Journal of Medical Sciences*, 1968, vol. 4, p. 276.)

IT is well known that certain rather rare hereditary diseases are less rare among Jews than among Gentiles, and that some other diseases, not definitely hereditary, are likewise commoner in Jews. On the other hand some diseases are distinctly rarer in Jews than in Gentiles, and others vary in frequency between the main groups of Jews.

The literature of this subject is large and scattered, and we cannot claim to have seen more than a small fraction of it. All that can be done here is to give a very short summary of the main topics, based largely on readily accessible secondary sources. One of the most convenient of these, and one which will be convenient also for consultation by our readers, is the volume on *The Jews*, edited by Shiloh and Selavan (1973) in the series 'Ethnic groups of America; their morbidity, mortality and behaviour disorders'. This book consists of a series of reprints of some of the more important papers on the subject, together with editorial comments. Each of the included papers is listed separately, with its original reference, in the bibliography of the present book. Patai and Wing (1975) also have a chapter listing the main 'Jewish' diseases, and discussing their implications as regards the genetics of the various Jewish communities. Disease names have also been compared with the International Classification of Diseases (World Health Organization 1967) and both names and accounts of genetics with the catalogue of McKusick (1966), which also supplied some further names of diseases for inclusion here.

DISEASES ESPECIALLY AFFECTING ASHKENAZI JEWS

A large part of the literature consists of studies of the incidence of diseases among Ashkenazi Jews, compared with their incidence in other Europeans. Only in a relatively few cases are there comparable data for the other major Jewish communities, but studies now in progress in Israel are doing much to redress the balance.

A most useful brief summary of the whole situation, entitled 'Jews, genetics and disease', is that of Post (1965), who lists eight hereditary or familial diseases which are notably commoner among Ashkenazi Jews than among other Europeans. These are: dystonia musculorum deformans, familial autonomic dysfunction, Gaucher's disease, infantile amaurotic idiocy, lipid histiocytosis, pemphigus vulgaris, essential pentosuria, and polycythaemia vera.

Dystonia musculorum deformans (torsion dystonia) is characterized by choreic movements and tremors. It is usually inherited as a dominant, but with variable expressivity. It is found especially, but by no means exclusively, in Ashkenazi Jews.

Familial autonomic dysfunction (familial dysautonomia, Riley–Day Syndrome) is characterized by a variety of neurological symptoms, of which the most important clinically is difficulty in feeding and swallowing, leading to regurgitation and recurrent aspiration bronchopneumonia; at least one-fourth of the patients die before the tenth year (McKusick *et al.* 1967). The condition is inherited as an autosomal recessive and is almost entirely confined to Ashkenazi Jews, in whom the incidence is between 1 in 10 000 and 1 in 20 000 births, indicating a gene frequency slightly below 1 per cent. McKusick *et al.* suggest that the basic defect is lack of an enzyme involved in the metabolism of acetylcholine.

Gaucher's disease (familial splenic anaemia) of the chronic adult type is one of the lipoidoses, showing accumulation of sphingolipids in the tissues, and gross enlargement of the spleen. It is inherited as a recessive in most pedigrees but as a dominant in some. It seems likely that the fully developed disease occurs only in homozygotes but that heterozygotes may show mild symptoms. Groen (1964) has found the disease to become more and more severe in successive generations. His observations are not questioned, but that this should appear to be a general rule is probably an accident of the methods of ascertainment.

Infantile amaurotic idiocy (Tay–Sachs disease) is the best known and one of the commonest of the hereditary diseases which predominantly affect Ashkenazi Jews. It has a well-defined recessive mode of inheritance, the heterozygous parents being clinically normal. It is a disease of lipid storage due to a lack of the enzyme hexosaminidase A (N-acetyl β-D-hexosaminidase A). As a result the sphingolipid ganglioside accumulates in the cytoplasm of the neurons of the brain. Degeneration of cerebral function begins soon after birth, and death occurs usually during the first or second year of life. Myrianthopoulos and Aronsohn (1966) have suggested that the disease persists from one generation to the next because the heterozygous carriers are more fertile than normal persons.

Lipid histiocytosis (Niemann–Pick Disease), like Tay–Sachs and Gaucher diseases, is characterized by an accumulation of sphingolipid in the spleen which may reach very large dimensions. Two distinct types of the disease are described, an acute fatal infantile type and an adult chronic one, of which only the latter occurs in Jews. This suggests that the two types are genetically separate. The adult type appears to behave as an autosomal recessive. The incidence among Ashkenazi Jews (Fuhrmann 1967) is said to be between 1 in 2000 and 1 in 3500, implying the very high gene frequency of about 2 per cent. It is virtually absent in other Jewish communities.

Pemphigus vulgaris is a skin disease, mainly of adult onset, characterized by giant bullae, with a highly characteristic histological picture, and the presence of anti-epithelial antibodies in the serum. Before the introduction of steroid therapy it had a high mortality, and steroids, in the large doses needed, have tended to produce severe complications of their own. Gold (sodium aurothiomalate) also appears to have some therapeutic effect (as well as its own side-effects).

The diesase is a rare one. It appears to be much commoner among Ashkenazi Jews than in any other population, but occurs also in non-Jewish Caucasoids and in Negroes. It has a tendency to familial occurrence but no inheritance pattern has emerged. The main clue to its heredity is in fact supplied by the observations of Krain et al. (1973) on its associations with histocompatibility antigens.

In Ashkenazi Jews these authors found a very strong association with HL-A10, the antigen being present in 61 per cent of 43 pemphigus cases (71 per cent of the females; 45 per cent of the males) but in only 20 per cent of controls. This might be due either to the HL-A10 antigen itself being part of the cause of pemphigus, or alternatively to a pemphigus-susceptibility gene forming part of the closely linked HLA set of genes on chromosome 6. In the latter case there would have to be linkage disequilibrium between the HL-A10 gene and that causing the disease, i.e. the two genes would be present together on one chromosome more often than would be predicted by complete crossing-over equilibrium.

In non-Jewish white patients there was a much less marked association, with 20 per cent (of only 15 patients) having the antigen, compared with 11 per cent in controls. These small numbers do not provide a very strong basis for theorizing, but they point to linkage disequilibrium (rather than a direct effect of the HLA genes) as the cause of the observed association, the degree of disequilibrium being much more marked in Ashkenazi Jews than in non-Jewish whites. It is likely that the causation of the disease is polygenic, a suceptibility gene on chromosome 6 being one of several genes involved.

Katz et al. (1973) studied a much smaller series of 18 patients, of whom 7 are said to have been 'either of Jewish or Mediterranean origin or both'. Their HLA observations though showing only a small rise in HL-A10 frequency, do not differ significantly from those of Krain et al.

In a survey of the origins of patients treated by them in Israel, Ziprkowski and Millet (1964) confirm the exceptionally high incidence of the disease among Ashkenazi Jews. However, in addition to 33 Ashkenazim there were 13 classified as Sephardim (from Turkey, Syria, Kurdistan, Iraq, Egypt, Libya, Tunisia, and Morocco) and 2 Yemenites. Thus the disease, though predominantly characteristic of Ashkenazim, is to be regarded as occurring in Jews of all origins more frequently than in non-Jews.

The total numbers of cases described by the above-mentioned authors are, however, still small, and further statistics are needed, both as to the incidence of pemphigus vulgaris in different communities, Ashkenazi, other Jewish, non-Jewish whites, Mongoloids, and Negroes, and also as to the histocompatibility associations of the disease in each community. Such studies of this rather rare disease might yield valuable information on the population dynamics of linkage disequilibrium in general.

Essential pentosuria (1-xylulosuria) is not a disease but a biochemical curiosity due to hereditary lack of an enzyme; it is recessively inherited. Its chief disadvantage is that it is liable to be mistaken for diabetes mellitus in medical examinations for insurance purposes. It is found mainly in Ashkenazi Jews from Poland and Russia, but is also not uncommon in Lebanese Arabs (Khachadurian 1962).

Primary adult lactase deficiency, also, cannot be regarded as a disease, for it affects more than half the human race. It can, however, in the presence of an inappropriate milk diet, give rise to digestive disturbances. It is probably inherited as a recessive condition. As shown in Table 33, it occurs in Jews of various communities with frequencies ranging from 54 to 85 per cent, and averaging 66 per cent. It is noteworthy that the Ashkenazim have the rather high frequency of 64 per cent, very much above other Europeans, though not as high as Cypriots, Arabs, or Sephardi or Oriental Jews. Before any firm conclusion can be drawn there is a need for more data on non-Jews from all parts of Europe, and from the different countries of north Africa.

Polycythaemia vera (erythraemia) is characterized by an excessive concentration of red cells in the blood, giving rise to thickening of the blood and hypertension. It appears to be hereditary as a dominant condition but has some of the characteristics of a malignant disease (cf. leukaemia). Krikler (1970), unlike Post, does not consider it to have a particularly high incidence in Jews.

Two other very rare recessively expressed conditions are found in Jews in America, who are most likely to be Ashkenazim. Isselbacher et al. (1964) describe a recessive abetalipoproteinaemia which manifests itself by a syndrome comprising celiac disease (see p. 55), acanthocytosis (burr-cell shape) of the red cells, and progressive ataxic neuropathy. This occurs in two families, of which one is stated to be Jewish. Bloom's syndrome, dwarfism with excessive skin sensitivity to sunlight, is described in three families, two of which are stated to be Jewish (Bloom 1966).

Paronychia congenita, a dominantly expressed condition characterized by hyperkeratosis of nails and skin, is perhaps commoner in Ashkenazi Jews than in other communities (McKusick 1966).

Recessive glycinuria (excess of the amino-acid glycine in the urine) has been found in a Bulgarian Jewish family. It appears to be distinct from the much more serious glycinaemia (McKusick 1966).

Dominant absence of upper canine teeth has been described in a German Jewish family (Grüneberg 1936), and recessive spongy degeneration of the central nervous system in Jews of Vilna (Efron, personal communication to McKusick 1966).

DISEASES PECULIAR TO OTHER JEWISH COMMUNITIES

The Dubin–Johnson syndrome (hyperbilirubinaemia, chronic idiopathic jaundice) (Dubin and Johnson 1954) is characterized by a chronic or intermittent hyperbilirubinaemia, accompanied by a deposit of dark pigment in the liver cells. It is usually also marked by a deficiency of clotting factor VII (a deficiency which also occurs by itself as a rare genetic abnormality: Seligsohn et al. 1970). The syndrome appears to be inherited as an autosomal recessive, but it is possible that heterozygotes may sometimes show mild symptoms. McKusick (1966), however, classified it as a dominant condition, though he lists several other hyperbilirubinaemias as recessive. The disease is found in a wide variety of populations, Jewish and other, but is especially prevalent in the Jews of Iran, particularly those of Isfahan (Shani *et al.* 1970).

Wolman's disease (familial xanthomatosis) (Wolman *et al.* 1961) was first found in an Iranian Jewish family, but there is no strong evidence that it is particularly characteristic of this community. Xanthomatous deposits occur in various organs; the condition is of recessive inheritance.

Atypical pseudocholinesterase in the blood plasma is under natural conditions a harmless curiosity, but homozygotes for any of the weak variants, when subjected to relaxation anaesthesia by means of succinyl-choline, remain relaxed and fail to resume spontaneous breathing for dangerously long periods. The least rare of the abnormal genes, E_1^u, is present in most populations with a frequency of between 1 and 2 per cent, but the incidence is higher than this in several Oriental Jewish populations, reaching 7·5 per cent in the Jews of Iran, 4·7 per cent in those of Iraq, and 3·6 per cent in Yemenite Jews (Sheba 1971; data available in a form suitable for reproduction in the tables of this book do not separate Iraqi and Iranian Jews).

Kurdish Jews, who may come from either Iran or Iraq (more rarely from Turkey), have very high frequencies of glucose-6-phosphate dehydrogenase deficiency. This must be regarded as a diathesis rather than a definite disease, and it is discussed elsewhere (p. 9).

Like this deficiency, thalassaemia is prevalent throughout most of the Mediterraean area and, like it, is thought to give some protection from malaria. Both varieties of thalassaemia, α and β (see p. 12) have rather high frequencies in the Kurdish Jews.

A recessive syndrome consisting of nerve deafness and infantile renal tubular acidosis has also been described in Kurdish Jews (Cohen *et al.* 1973).

A surprisingly large number of very rare hereditary diseases and syndromes, mostly recessive, show relatively high frequencies in the Yemenite Jews. However, unlike the characteristic diseases of the Ashkenazi Jews, few cases of them have found their way into the hospitals of western Europe and America, and the diseases of the Yemenites are thus less well known outside Israel.

Among such recessive, or probably recessive, conditions are familial neonatal hyperbilirubinaemia (Sheba, personal communication to McKusick 1966), and a condition diagnosed by Nisenbaum *et al.* (1965) as 'Pelizaeus–Merzbacher disease, infantile acute type'. McKusick, however, does not agree with the name which they apply to this neuropathological condition. Yemenite Jews are also the main sufferers from a blood coagulation defect, deficiency of plasma thromboplastin antecedent (Factor XI) (Rosenthal 1964). As interpreted by McKusick (1966), they also share with the Maori of New Zealand a rather high incidence of recessive cystic disease of the lung (Racz *et al.* 1964; Hinds 1958).

The Yemenite Jews have a rather high incidence of (dominant) benign familial neutropenia (lack of neutrophil leucocytes) (Feimaro and Alkan 1968).

Yemenite Jews also have the highest known incidence (about 9 per thousand) of recessive deficiency of peroxidase and phospholipid in the eosinophil granulocytes of the blood. Lower frequencies are found in north Africa and Iraqi-Persian Jews, and the condition appears to be totally absent

among the Sephardim of Europe and the Ashkenazim (Joshua *et al.* 1970).

Yemenite and north African Jews show a much higher incidence than other Jewish communities, of childhood celiac disease, a condition which undoubtedly has a genetic component in its aetiology (Lasch *et al.* 1968).

The Yemenite Jews have the highest incidence of phenylketonuria known among Jews. This is a recessive condition due to lack of the liver enzyme phenylalanine hydroxylase, which catalyses an essential step in the metabolism of phenylalanine, an important amino-acid constituent of most but not all dietary proteins. In its absence metabolism stops at the stage of phenylpyruvic acid, the accumulation of which causes irreparable damage to the developing brain. Some damage has already occurred by the end of pregnancy but if the condition is recognized at birth, and the infant put on a phenylalanine-free diet, mental development is almost normal.

The disease is of world-wide distribution but the highest frequencies occur in north-west Europe, and especially in Scotland, Ireland, and Iceland, with frequencies of 29, 25, and 13 per hundred thousand births (Saugstad 1975). It occurs also, with somewhat lower frequencies, in England and the continental Scandinavian countries. In Norway its overall frequency is 6 per hundred thousand but this is largely concentrated in areas which appear to have been settled by a Celtic population. The parents of affected Norwegian infants show raised frequencies of Rh-negatives, of Kell (K) positives, and of the phosphoglucomutase type $PGM_1 1$; this is consistent with a Celtic origin and persistent stratification of the population. It has been suggested that this deleterious gene persists because of some selective advantage enjoyed by the heterozygous parents.

Among Jews the frequency of this condition is in general very low and it appears to be completely absent among the Ashkenazim, but among the Yemenite Jews it is relatively high, possibly as high as 15 per hundred thousand, a level comparable with that in north-west Europe. No obvious anthropological explanation can be given, and this may be an example of genetic drift.

A number of diseases are said to be particularly prevalent among north African Jews—some in particular countries, some in the whole of north Africa.

There are however few if any diseases particularly attributed to Egyptian Jews or to Algerian Jews. It is not clear whether Jews in the two countries concerned are particularly healthy or whether, more probably, adequate data are lacking. It may well be that a study of adequate numbers would reveal some interesting clines of congenital disease prevalence, extending along the whole north African coast.

Familial Mediterranean fever (recurrent polyserositis) is characterized by recurrent attacks of fever with pain, usually in the abdomen but sometimes in chest, joints, or skin. Pathologically it is characterized by hyperaemia and nonbacterial inflammation of the serous membranes. Amyloidosis may supervene and slowly lead to fatal kidney failure. The condition is transmitted as a recessive autosomal with incomplete penetrance.

It is found mainly in Sephardic Jews, especially those from Libya, in Jews from Iraq, and in Armenians.

Congenital deafness (deaf-mutism) also is said to be especially prevalent in north African Jews (as it is also in the Samaritans).

Two related and highly consanguineous Moroccan Sephardic Jewish families (Ziprkowski and Adam 1964) show recessive deafness, combined in most but not all cases with albinism. A syndrome of X-linked deafness and partial albinism in Jews has also been described by Ziprkowski *et al.* (1962) and by Margolis (1962). It would be interesting to know whether either of these syndromes is related to the well-known combination of deafness and whiteness in cats.

One of the glycogen storage diseases, (amylo-1,6-glucosidase deficiency, or limit dextrinosis) is stated by Krikler (1970) to be an autosomal recessive condition with a particularly high frequency in north African Jews.

Other diseases found especially in north African Jews are cystinuria (Libya), vitamin B_{12} malabsorption (Libya and Tunisia), and Ataxia-telangiectasia (Morocco).

The Creutzfeldt–Jacob syndrome is a chronic generalized degenerative condition affecting the central nervous system. It occurs especially in Libyan Jews (Kahana *et al.* 1964), but appears to be due to a 'slow' virus and not to a human gene. According to Herzberg *et al.* (1974) and Alter (1974), it is probably due to the eating of undercooked brain, spinal cord, and eyeballs of sheep infected with the scrapie virus. In its mode of transmission the disease thus resembles kuru in New Guinea, formerly thought to be genetic, but now known to be due to a virus transmitted by the ceremonial cannibalistic eating of infected human brains.

There are a great many diseases which tend to be familial, but which are largely dependent upon the environment. Such are, for instance, the diseases of affluence, including many cardiovascular diseases and diabetes mellitus. While these diseases, for which there are good statistics, differ in incidence from one Jewish community to another, the differences are likely to be very largely environmental, and not likely to throw any clear light on genetic differences between the various Jewish communities.

The same is probably true of the many forms of cancer. For these the statistics of the state of Israel are particularly complete. While a full analysis of this great wealth of material would probably reveal congenital differences in susceptibility between the main Jewish communities, it is likely to be even more valuable in indicating the environmental factors which are undoubtedly highly important in the causation of cancer.

DIABETES MELLITUS

It has been claimed that diabetes mellitus is particularly common among Jews, but it is difficult to ascertain how far this is a cultural rather than a genetic phenomenon.

Krikler (1969) has studied the incidence of diabetes among Jews in Rhodesia, and finds it much higher in Sephardim than in Ashkenazim. Most of the Sephardim in Rhodesia originate in the Greek islands of Rhodes and Cos, and the adjacent Turkish mainland. Krikler attributes the high frequency to the great frequency of consanguineous marriages among these Sephardim, and perhaps also to selection for diabetes (the 'Thrifty genotype' of Neel, 1962) under the condition of poverty prevailing in Rhodes. It is noteworthy

that, among the cases classified for age of onset, all those in Sephardim are of adult onset. It is probable that adult-onset diabetes is more strongly hereditary than the juvenile type, and the 'thrifty genotype', though inborn, does not usually manifest itself pathologically until adult life.

In a study of the genetics of different Jewish populations, the data of greatest interest are those relating to diseases which are due mainly to genes which are abnormal and the expression of which is unaffected or little affected by the environment. As we have seen, the Ashkenazim are particularly affected by a number of such conditions, most of which are due to autosomal genes with recessive expression.

There are two main ways (not necessarily mutually exclusive) in which such harmful genes can come to have the rather high frequencies which have been found. One is through random founder effects and genetic drift. The other is through natural selection, which in this situation implies that while homozygotes for the abnormal gene are at a severe disadvantage, heterozygotes enjoy certain selective advantages in the particular environment to which the communities have been exposed. While heterozygote advantage has certainly been involved for some of the conditions mentioned, and especially for thalassaemia, in most cases the evidence is very slender and founder effect and drift constitute the more likely explanation.

If the same disease genes were found in all, or nearly all, Jewish communities, then it would be possible to attribute their incidence to a common founder effect in the original Palestinian Jewish community. But the diseases present differ from one community to another. Thus founder effects (possibly following chance mutations) would need each to be peculiar to one of the major Jewish communities. However, as we have seen in the historical sections of previous chapters, most of these major communities (except perhaps the Babylonian one) appear to have been built up by long-continuing trickles of population from Palestine, or from secondary sources of dispersal. On the other hand, many of the communities have been cut down at times to very small numbers by persecution and extermination, or by epidemics, and have then increased considerably to something like their original numbers. These are the conditions under which random genetic drift (at times of minimum population) are likely to have occurred.

However, it is especially in the Ashkenazim that these particular disease genes are to be found, and this community has, for much of its existence, been quite numerous. We must therefore be on our guard against arguing in a circle. We do not really know the minimum numbers to which the Ashkenazim have been cut down at the darkest periods of their history. Certainly the half-million or so who survived the Nazi holocaust were far too many to have shown appreciable genetic drift.

We cannot, on present evidence, reach any firm conclusion, but in my opinion genetic drift, at times of bottlenecks in the numbers of one or more main populations who ultimately made up the Ashkenazim, is the most likely cause for the disease gene frequencies which we now find in them.

12

SOME GENERAL CONCLUSIONS

THE published information on the genetics of the Jews is highly heterogeneous and, therefore, difficult to sum up. In particular, and apart from some recent work, each blood-group survey has been planned without much reference to those done previously, and slight differences in the choice of systems studied, and in the particular tests carried out, make it difficult to combine the results statistically. We have therefore, in general, considered each blood-group system separately. Moreover, since very much more information is available on the ABO system than on any other, a great many of our conclusions have been based mainly or entirely on this system, though we still do not know, despite much investigation, how far the phenotypes of this system are subject to natural selection, such as might result from particular types being at an advantage or a disadvantage in certain environments (e.g. climatic or microbiological).

It may be said that, in general, blood-group data are readily correlated with the known facts of history, and that they support the relative homogeneity of the main historical Jewish communities and their distinctness from one another. Here we shall stress mainly those sets of observations where the blood-group data give positive support to accepted history, those where there appears to be important disagreement, and those where new historical conclusions can be drawn from the blood groups.

It is difficult to draw any conclusions from the study of living Jewish communities as to the genetic composition of the original Israelites of Palestine, and the nearer to the present day is the origin of any particular community the more we can learn from its blood groups.

There are however, certain generalizations that can be applied to all or nearly all Jewish communities. Each major community as a whole bears some resemblance to the indigenous peoples of the region where it first developed and, within each community there is some relation between the compositions of the separate Jewish sub-communities (national etc.) and those of the peoples among whom they have recently lived. Nearly all Jewish communities show a substantial proportion of African Negroid marker genes, such as to imply a total Negroid admixture of the order of 5 to 10 per cent. These admixtures are readily explained by slavery and concubinage, as set out in some detail by Patai and Wing (1975).

The Jews of Kurdistan, Persia, and lands to the north and east of these show a considerable scatter of gene frequencies, but fit fairly well into the regional non-Jewish picture. Their traditions suggest that they are descended, at least in part, from the 'lost tribes', both those from east of the Jordan, taken into captivity by Tiglath-Pileser, and those deported from northern Palestine by Sargon II. Their rather low frequency of African marker genes suggests that they have long remained genetically separate from the more southerly Jews of the Babylonian captivity and of the Dispersion into north Africa and Europe.

The Iraqi Jews have a historic claim to be the direct and more or less unmixed descendants of the Jews taken into captivity from Jerusalem and the surrounding area by Nebuchadnezzar. They too, however, do not depart widely, in their present genetic composition, from the surrounding non-Jewish peoples. The relations between Jewish and non-Jewish ABO frequencies in south-west Asia are summarized in Fig. 2.

One of the clearest facts to emerge from blood-group studies is the distinctness of the Yemenite Jews from the rest of the Oriental Jews, and their close resemblance to the Arabs of the Arabian peninsula (who are themselves distinct from all other cultural Arabs). There can be little doubt that substantial numbers of Jews migrated to the south at about the time of the destruction of the second Temple, but they seem to have become merged genetically into the more numerous Arab community. Nevertheless for a time those who had remained Jews by religion prevailed culturally and set up a Jewish Himyarite kingdom in southern Arabia. With the coming of Islam most of the inhabitants of the area embraced that religion, but there are no clear indications of any genetic factor dividing those who did so from those who remained Jews.

Jews had been moving from Palestine into Egypt and thence into the other countries of north Africa since long before the Christian era, so that, by the time of Mohammed, there undoubtedly were Jews in all the north African coastlands. Besides Jews of Palestinian descent, these must have included many who were descended from indigenous pagan peoples who had adopted the Jewish religion. The Jews of north Africa, together with those of Mesopotamia, supplied the immigrants who gave rise to the very large body of Jews who lived in Spain during the many hundreds of years of the Muslim occupation of that country. With the expulsion of both Muslims and Jews from Spain in A.D. 1492 a new stratum was superimposed on the north African Jewish communities, but it is now very difficult to distinguish the two component strata from one another, at least in the coastlands. There are however distinct differences between the gene frequencies (at least for ABO) of the Jews of the different independent nations which to-day constitute north Africa, and these to some extent run parallel to the differences between their non-Jewish populations.

In the countries of southern Europe and the Mediterranean coastlands of the Near East, there were undoubtedly Jewish communities in the Middle Ages, but outside Spain and Italy we have no knowledge of their having been very numerous. Thus the Sephardim who fled from Spain in 1492 and the following years seem to have come to constitute the majority of the Jewish population of these countries, and they do indeed show a certain genetic uniformity throughout the region. The difficulty here is to account for their original genetic composition, a problem discussed at length on pp. 41–43.

One of the major geographical features of Europe is the series of mountain ranges, from the Alps to the Caucasus, which separates the great plains to the north from the Mediterranean region to the south. It would be easy to account arithmetically for the genetic composition of the

Ashkenazim as being derived from the Jewish populations who, in the early Middle Ages, lived just south of those mountains, together with some admixture of the indigenous peoples of the northern plains.

For reasons stated in Chapter 10, it was, however, considered necessary to devote particular attention to Koestler's theory of the derivation of the Ashkenazim from the Khazars, a Turkish nation living near the Caspian Sea, and converted to Judaism; the conclusion was reached that the Khazars probably made only a small genetic contribution, if any, to the Ashkenazim, but that they may have been the main ancestors of the Karaites of the Soviet Union (Crimea and Lithuania) and of the orthodox Krimchaks of the Crimea.

Whereas in this book we have tried to analyse the unwieldy total body of genetic data, Cavalli-Sforza and Carmelli (1977) have approached the problem in another way, selecting only data which were amenable to a computerized multivariate analysis. They have very kindly allowed us to see their unpublished results.

Their method, while much more completely objective than our own, can be applied only to a set of populations all of which have been tested for the whole of a particular set of genes. For this purpose they have selected 14 Jewish populations and 25 non-Jewish ones, all of which have been tested for the gene products of the ABO, MN, Rh, Haptoglobin, and Gc systems. Subject to the limitations imposed by the method, their selection of populations appears appropriate for all the Jewish populations and most of the non-Jewish ones, but the indigenous Indian one does not appear to be fully representative, and in north Africa the Jews (mostly from the coastal towns and largely Sephardic) should have been compared with coastal Caucasoid Arabs and not with the largely Negroid populations of the desert fringe chosen in the case of Algeria and Libya, a choice which has certainly affected the results of the analysis. Nevertheless they show that the method correctly classifies most Jewish populations, despite their long history of dispersion,

as belonging to the Middle East (or more precisely, Near East) region, and only a few were instead classified in the geographic area in which they recently lived, indicating admixture with local populations. Direct estimates of such admixture were obtained; they were maximum for an Ashkenazi group, usually low or inconclusive for most others. The analysis by discriminant functions has also shown a considerable scatter of gene frequencies (their linear transformations) most probably due to drift.

These conclusions are very similar to our own reached by less precise but more widely ranging analysis.

It is important that in planning future population genetic studies, not only of Jews, but of all peoples, there should be some international agreement between laboratories as to a minimum set of genetic characters for which all populations should be tested, so as to facilitate multivariate analysis covering the largest possible range both of populations and of characters.

As we have seen in Chapter 11, each major Jewish community carries the genes for a number of congenital diseases. These are mostly of recessive expression, with gene frequencies of the order of one per cent and disease frequencies around one in ten thousand. The diseases are almost totally different in each community, and have been best studied in the Ashkenazim. The marked differences between the communities in this particular respect suggests that the genes concerned entered the various communities, or resulted from mutation within them, after the times when the communities concerned became genetically separate from one another. With a very few possible exceptions, these genes seem to be totally deleterious without, for instance, any compensating benefit in heterozygotes. Thus their present frequencies which, for harmful genes, are rather high, are likely to have arisen by genetic drift in very small communities which subsequently multiplied rapidly. It is thus probable that, within the last thousand years or less, the communities concerned have passed through a low-population bottle-neck such as would favour drift. In the case of a community such as the Ashkenazim, which may have originated by the merging of a few separate communities, a particular harmful gene may have been contributed by one only of these, provided that the drift and a subsequent considerable amount of expansion took place before it merged with any other large community.

The distribution of congenital diseases in Jews thus supports the view that small communities of them may have multiplied rapidly at certain periods and that, for instance, the large numbers of Jews who, according to some authors, suddenly appeared in Poland in the Middle Ages, were not necessarily the descendants of a large nation such as the Khazars.

TABLES

INTRODUCTION

EACH table is specified by one, two, or three numbers: the first or sole number refers to the genetical system concerned:

1–10	Major blood group systems
11–12	Other blood group systems
13–19	Plasma protein systems
20–30	Red-cell enzyme systems
31–33	Other systems of genetic markers

Where, within a given system, several testing routines, with differing degrees of genetic discrimination, are possible, these are distinguished by a second number. However, some population samples, upon which tests for the D^u antigen were performed with negative results, are, for the sake of compactness, included in Table 4.7, with the note 'No D^u' and are not included in Table 4.8, which lists separately phenotypes containing D^u. For each of three systems, moreover, (ABO, MN, and Rhesus), we have included *all* our data in the table giving the results of the simplest testing routine within the system. These three tables, thus, allow comparisons to be made at the most basic level. Against each entry simplified in this way, there is given, after the authors' names, the number of the Table in which the complete data are to be found.

For one system only, glucose-6-phosphate dehydrogenase, a third number has been necessary. This is because the system is sex (X) linked, and the sexes of the subjects therefore provide genetical information additional to that conveyed by the results of the tests.

The arrangement of the data is the same in each table, showing in successive columns, or groups of columns, for each set of data:

(1) Place
(2) Population
(3) Authors
(4) Number of individuals tested
(5) Number observed for each phenotype
(6) Gene frequencies
(7) χ^2 or other criterion for the goodness of fit of the expected phenotype frequencies with those observed.

(1) The geographical location shown *always* refers to the origin of the population as stated by the authors, and *not* to where it was sampled. When only that location is shown, this means that the population was sampled in Israel (as it was in the great majority of cases). The sign of a dagger (†) indicates that the population was sampled in the stated country of origin; the place of sampling is shown when the latter was neither Israel nor the place of origin. The names Yemen and Aden, as used in the tables, are those by which the territories were known at the time when the tests were carried out. Yemen is now known as the Yemen Arab Republic, and Aden forms part of the Peoples' Democratic Republic of Yemen.

(2) All the populations are, of course, Jewish (with the sole exception of one said to include 'half-Jews'), but this column gives all further relevant information provided by the authors.

(4) When the number of individuals tested is followed by the symbol (a), the published information has been supplemented by a personal communication from the authors.

(5) Where numbers observed are not stated by the authors, we have calculated them where possible from the published frequencies; a (c) following the total indicates that this was done. Where phenotype numbers are not shown, but only gene frequencies, this means that only the latter were published, and we have been unable to obtain information on phenotype numbers.

(6) Gene frequencies were calculated by the methods described by us in *The distribution of the human blood groups* (Mourant *et al.* 1976). In certain cases, especially those of blood-group systems where only one testing serum has been used, there are only two phenotypes distinguishable, and two genes involved; hence there are no degrees of freedom, and the frequencies of the genes are uniquely determined by those of the phenotypes. In all other cases the gene frequency estimates are maximum likelihood ones.

(7) The fit between observed phenotype numbers and the expected ones, calculated from gene frequencies, gives an indication of the reliability of the data. For this reason we show the values of either D/σ or χ^2 for goodness of fit. In the case of the Rhesus system, however, where the usefulness of χ^2 may be affected by too small expected numbers, the latter have been quoted *in extenso*.

TABLE 1.1
ABO SYSTEM: TESTS WITH ANTI-A AND -B

Place	Population	Authors		Number	O	A	B	AB	p	q	r	D/σ
SOUTH-WEST ASIA												
ISRAEL		Bar-Shany 1974		4205	1398	1631	853	323	26·89	15·15	57·96	+1·35
ISRAEL		Gurevitch 1940		640(c)	238	252	106	44	26·66	12·49	60·85	−0·25
ISRAEL		Dressler 1951		12 000(c)	4301	4920	2056	723	27·30	12·38	60·32	+3·85
ISRAEL	Born there	Stark & Henner 1953		633	237	237	113	46	25·60	13·45	60·95	−0·46
ISRAEL	Born there	Levene 1975	1·2	1403	470	507	312	114	25·37	16·57	58·06	
ISRAEL	Parents born there; sampled in Marseille	Lévy et al. 1967		40	17	6	12	5	14·54	23·66	61·80	−1·64
ISRAEL	Ashkenazi	Younovitch 1933b		1500	510	616	272	102	27·87	13·39	58·74	+1·19
ISRAEL	Ashkenazi	Gurevitch 1940		2662(c)	969	1065	442	186	27·18	12·58	60·24	−0·37
ISRAEL	Ashkenazi	Gurevitch et al. 1951		946	361	390	137	58	27·43	10·89	61·68	−0·24
ISRAEL	Ashkenazi	Margolis et al. 1960		465	177	187	74	27	26·57	11·54	61·89	−0·35
ISRAEL	Sephardi	Younovitch 1933		400	138	135	103	24	22·55	17·52	59·93	+1·69
ISRAEL	Sephardi	Gurevitch 1940		1107(c)	447	343	211	106	22·68	15·37	61·95	−4·02
ISRAEL	Sephardi	Gurevitch et al. 1951		252	102	81	47	22	22·93	14·67	62·40	−1·49
ISRAEL	Sephardi	Margolis et al. 1960		200	59	90	33	18	32·15	13·67	54·18	−0·14
ISRAEL	Oriental	Gurevitch 1940		451(c)	151	146	100	54	25·13	18·65	56·22	−2·28
ISRAEL	Oriental	Gurevitch et al. 1951		137	50	46	26	15	25·23	16·11	58·66	−1·44
ISRAEL	Oriental	Stark & Henner 1953		94	27	36	22	9	27·87	18·18	53·95	+0·22
ISRAEL, Nablus	Samaritans	Parr 1931		83	42	27	8	6	22·12	8·72	69·16	−1·83
ISRAEL, Nablus, Jaffa	Samaritans	Bonné-Tamir 1975	1·2	124*	66	43	14	1	19·76	6·28	73·96	
ISRAEL, Jaffa	Samaritans	Bonné 1966	1·2	132	89	27	15	1	11·26	6·27	82·47	
ISRAEL, Nablus	Samaritans, Levi	Younovitch 1938[1]		41	20	18	3		25·27	3·75	70·98	+1·04
ISRAEL, Nablus	Samaritans, Levi	Ikin et al. 1963	1·2	33(a)	8	20	4	1	40·14	8·01	51·85	
ISRAEL, Nablus	Samaritans, Ephraim	Younovitch 1938[1]		68	54	10	4		7·66	2·99	89·35	+0·59
ISRAEL, Nablus	Samaritans, Manasseh	Younovitch 1938[1]		70	36	17	13	4	16·25	12·92	70·83	−0·72
ISRAEL, Nablus	Samaritans, Ephraim (44)	Ikin et al. 1963	·1·2	49(a)	28	14	6	1	16·69	7·43	75·88	
LEBANON, Beirut†		Parr 1931		181(c)	52	62	35	32	29·92	20·13	49·95	−2·85
LEBANON and SYRIA		Stark & Henner 1953		64	20	23	14	7	26·97	17·94	55·09	−0·42
LEBANON and SYRIA		Levene 1975	1·2	204	72	77	41	14	25·70	14·57	59·73	
SYRIA, Aleppo†		Altounyan 1928		172	66	58	34	14	23·66	15·03	61·31	−0·62
SYRIA, Aleppo, Damascus		Younovitch 1933b		104	40	32	23	9	22·06	16·72	61·22	−0·60
MESOPOTAMIA												
IRAQ	Babylonian	Younovitch 1933b		210	58	72	42	38	30·24	20·80	48·96	−3·56
IRAQ		Stark & Henner 1953		30	7	13	8	2	29·88	18·73	51·39	+1·00
IRAQ	Females	Silberstein & Goldstein 1958		308(c)	81	128	61	38	31·94	17·53	50·53	−0·79
IRAQ		Levene 1975		1147	256	449	305	137	30·22	21·72	48·06	
IRAQ		Bonné-Tamir 1976	1·2	184	42	81	37	24	34·53	18·20	47·27	
IRAQ, Baghdad†		Kennedy & MacFarlane 1936	1·2	73	30	21	16	6	20·50	16·32	63·18	−0·61
IRAQ, Baghdad†		Boyd & Boyd 1941		215(c)	58	80	44	33	30·66	19·59	49·75	−1·86
IRAQ, Baghdad		Gurevitch & Margolis 1954–5		162	40	62	39	21	30·09	20·60	49·31	−0·27

IRAQ, Hit	Karaites	Goldschmidt *et al.* 1976	1.2	72	10	22	34	6	22·53	34·58	42·89	
IRAQ	Kurdish	Younovitch 1933*b*		147	48	60	21	18	31·03	14·08	54·89	−1·84
IRAQ	Kurdish	Gurevitch *et al.* 1953–4		250	67	117	34	32	36·00	14·04	49·96	−1·79
IRAQ	Kurdish	Gurevitch & Margolis 1954–5		129	36	53	23	17	32·02	16·76	51·22	−1·12
IRAQ, N.W.	Kurdish	Tills *et al.* 1977	1.3	61	16	27	12	6	32·42	16·08	51·50	
IRAQ, S.E.	Kurdish	Tills *et al.* 1977	1.2	50	20	8	19	3	11·73	25·17	63·10	
IRAN	Kurdish	Tills *et al.* 1977	1.3	106	30	35	29	12	25·31	21·66	53·03	
IRAN		Younovitch 1933		436	145	142	109	40	23·65	18·84	57·51	−0·22
IRAN		Gurevitch *et al.* 1956		200(c)	58	75	51	16	26·37	18·59	55·04	+1·06
IRAN	Females	Silberstein & Goldstein 1958		225(c)	75	83	45	22	26·85	16·13	57·02	−0·72
IRAN		Levene 1975	1.2	154	51	53	37	13	24·40	17·84	57·76	
IRAN		Levene *et al.* 1977	1.2	159	51	44	51	13	20·04	22·79	57·17	
IRAN, Isfahan†		Kennedy & MacFarlane 1936		29	12	5	10	2	12·86	23·36	63·78	−0·24
IRAN	Sampled in U.S.S.R., Tbilisi	Semenskaya *et al.* 1937*a*		132	30	61	30	11	33·03	17·20	49·77	+1·41
IRAN and IRAQ		Kennedy & MacFarlane 1936		102	42	26	26	8	18·27	18·27	63·46	−0·56
IRAN and IRAQ		Bar-Shany 1974		4982	1303	1906	1151	623	29·72	19·68	50·60	−2·20
YEMEN		Younovitch 1932		1000	560	261	161	18	15·14	9·42	75·44	+2·26
YEMEN		Gurevitch 1940		98(c)	41	26	18	13	21·91	16·93	61·16	−2·58
YEMEN		Brzezinski *et al.* 1952		500	263	150	69	18	18·50	9·11	72·39	−0·33
YEMEN		Dreyfuss *et al.* 1952	1.2	104	46	45	10	3	26·60	6·47	66·92	
YEMEN		Stark & Henner 1953		178	91	56	24	7	19·60	9·12	71·28	−0·30
YEMEN	Females	Silberstein & Goldstein 1958		167	81	58	21	7	21·82	8·76	69·42	−0·29
YEMEN		Bar-Shany 1974		2309	1129	788	304	88	21·22	8·88	69·90	−0·13
YEMEN		Levene 1975	1.2	374	210	123	32	9	19·54	5·64	74·82	
YEMEN, N. and N.E.		Tills *et al.* 1977	1.2	65	38	23	4		19·71	3·14	77·15	
YEMEN, S.		Tills *et al.* 1977	1.2	92	57	20	10	5	14·50	8·43	77·07	
YEMEN, Habban		Bonné *et al.* 1970	1.3	595	249	118	181	47	14·93	21·34	63·73	
YEMEN, Saada, San'a, Damar, Beida, Aden		Bodmer *et al.* 1972	1.3	202(a)	108	65	22	7	19·74	7·45	72·81	
ADEN		Levene 1975	1.2	127	50	46	23	8	24·21	13·06	62·73	
INDIA												
		Bar-Shany 1974		404	138	109	123	34	19·64	21·82	58·54	+0·13
BOMBAY†	Baghdadi	Sirsat 1956	1.2	200	45	60	68	27	24·98	27·58	47·44	
BOMBAY†	Bene-Israel	Sirsat 1956	1.2	200	78	50	59	13	17·30	20·03	62·67	
KERALA, Cochin		Gurevitch *et al.* 1955		275	149	41	73	12	10·12	16·83	73·05	−1·00
KERALA, Cochin†	'Black'	MacFarlane 1937		106(c)	78	11	17		5·34	8·39	86·27	+1·05
KERALA, Cochin†	'White'	MacFarlane 1937		50(c)	9	31	10		40·02	11·02	48·96	+3·01
AFRICA												
ALGERIA, Oran†		Solal & Hanoun 1952		205	79	57	55	14	19·17	18·57	62·26	+0·19
ALGERIA, Oran†		Auzas 1957	1.2	1445	580	451	310	104	21·47	15·50	63·03	
ALGERIA	Parents born there; sampled in Marseille	Lévy *et al.* 1967		128	60	26	26	16	17·62	17·62	64·76	−3·39
ALGERIA		Bar-Shany 1974		643	240	210	138	55	23·24	16·28	60·48	−1·13
EGYPT		Levene 1975	1.2	536	180	187	121	48	25·08	17·23	57·69	
EGYPT	Karaites	Goldschmidt 1967	1.2	250					7·2	34·5	57·7	

*Data overlap with Ikin *et al.* 1963 and with Bonné 1966.
[1] Quoted by Genna 1938.

TABLE1.1 ABO SYSTEM: TESTS WITH ANTI-A AND -B (*cont.*)

Place	Population	Authors		Number	O	A	B	AB	p	q	r	D/σ
AFRICA (*cont.*)												
LIBYA	Females	Silberstein & Goldstein 1958		193(c)	53	79	37	24	31·49	17·18	51·33	−0·90
LIBYA		Bar-Shany 1974		1071	290	481	197	103	32·61	15·17	52·22	+0·38
LIBYA		Levene 1975	1.2	100	28	42	21	9	30·15	16·40	53·45	
LIBYA		Bonné-Tamir 1974	1.2	148	44	75	18	11	35·17	10·32	54·51	
LIBYA, Tripolitania		Gurevitch et al. 1955		200	75	65	50	10	21·09	16·45	62·46	+1·30
MOROCCO†		Gaud & Médioni 1948		89(c)	24	30	29	6	23·19	22·45	54·36	+1·30
MOROCCO†		Kossovitch 1953		730	269	264	145	52	24·70	14·55	60·75	+0·10
MOROCCO		Younovitch 1933b		160	56	53	46	5	20·48	17·74	61·78	+2·45
MOROCCO		Margolis et al. 1957		220(c)	83	73	44	20	23·87	15·69	60·44	−1·07
MOROCCO	Females	Silberstein & Goldstein 1958		1014(c)	374	290	276	74	19·96	19·11	60·93	+0·47
MOROCCO		Bar-Shany 1974		6089	2334	1741	1559	455	20·03	18·18	61·79	−0·71
MOROCCO		Bonné-Tamir 1976	1.2	191	71	46	53	21	19·19	21·50	59·31	
MOROCCO		Levene 1975	1.2	426	159	108	128	31	17·97	20·85	61·18	
MOROCCO	Parents born there; sampled in Marseille	Lévy et al. 1967		60	26	17	11	6	21·19	15·14	63·67	−1·32
MOROCCO, Marrakech†		Mechali et al. 1957	1.2	192	77	60	39	16	22·25	15·42	62·33	
MOROCCO, Mogador†		Mechali et al. 1957	1.2	64	22	18	19	5	20·03	20·97	59·00	
MOROCCO, Rabat†		Leblanc 1946		856(c)	326	239	227	64	19·62	18·75	61·63	−0·17
MOROCCO, Rabat†		Mechali et al. 1957		526	171	184	131	40	24·34	17·93	57·73	+1·11
MOROCCO, Tafilalet†		Ikin et al. 1972	1.2	146	48	25	56	17	15·42	28·96	55·62	
MOROCCO, Tafilalet, Erfoud†		Mechali et al. 1957		200	69	42	73	16	15·74	25·51	58·75	+0·02
MOROCCO, Tafilalet, Gourrama†		Mechali et al. 1957		150	39	34	66	11	16·53	30·60	52·87	+1·43
MOROCCO, Tafilalet, Rich†		Mechali et al. 1957	1.2	150	47	30	61	12	15·20	28·39	56·41	
TUNISIA†		Caillon & Disdier 1930		200	82	62	31	25	24·33	14·85	60·82	−3·39
TUNISIA, Île de Djerba†		Moullec & Abdelmoula 1954		70	33	15	18	4	14·61	17·15	68·24	−0·31
TUNISIA, Île de Djerba, Hara-Kbira†		Ranque et al. 1964		111	33	47	22	9	29·72	15·16	55·12	+0·41
TUNISIA, Île de Djerba, Hara-Srira†		Ranque et al. 1964		60	23	16	17	4	18·39	19·41	62·20	+0·17
TUNISIA		Margolis et al. 1957		200	79	62	45	14	21·25	16·03	62·72	−0·13
TUNISIA	Females	Silberstein & Goldstein 1958	1.2	408(c)	149	144	80	35	24·98	15·20	59·82	−0·90
TUNISIA		Levene 1975		99	35	36	18	10	26·56	15·19	58·25	
TUNISIA		Bar-Shany 1974		1584	500	614	332	138	27·54	16·15	56·31	+0·32
TUNISIA	Parents born there; sampled in Marseille	Levy et al. 1967		49	15	18	9	7	29·52	17·64	52·84	−1·08
ETHIOPIA†	Falasha	Bat-Miriam 1962	1.2	152	65	58	22	7	24·37	10·06	65·57	
REPUBLIC OF SOUTH AFRICA, Cape†		Bronte-Stewart et al. 1962		1115	403	440	185	87	27·31	13·02	59·67	−1·10
EUROPE												
WESTERN	Females	Silberstein & Goldstein 1958		222(c)	93	83	36	10	23·84	10·99	65·17	+0·59
SOUTHERN and SOUTH-EASTERN		Bar-Shany 1974		4809	1756	1965	765	323	27·59	12·04	60·37	−0·27
		Bar-Shany 1974		2598	798	1145	432	223	31·17	13·51	55·32	−0·36
EASTERN†		after Routil 1933		5271	1803	2175	930	363	28·04	13·15	58·81	+1·65
EASTERN		Bar-Shany 1974		25 637	8892	10 760	4124	1861	28·74	12·44	58·82	−0·81

Country	Reference	Note	Code									
AUSTRIA	Stark & Henner 1953			361	137	142	72	10	24·20	12·24	63·56	+3·06
AUSTRIA	Levene 1975		1.2	187	59	96	22	10	34·34	8·99	56·67	
BULGARIA	Stark & Henner 1953	Sephardi		57	12	32	11	2	37·48	12·48	50·04	+2·01
BULGARIA	Levene 1975		1.2	283	72	131	52	28	33·93	15·35	50·72	
CZECHOSLOVAKIA†	Škaloud 1934[1]			144	34	72	32	6	33·09	14·55	52·36	+2·87
CZECHOSLOVAKIA, Bratislava†	Trokan 1929[2]			535(c)	186	190	118	41	24·64	16·19	59·17	+0·34
CZECHOSLOVAKIA	Stark & Henner 1953			239	96	89	37	17	25·30	11·97	62·73	-0·81
CZECHOSLOVAKIA	Levene 1975		1.2	380	161	140	56	23	24·35	10·97	64·68	
GERMANY, Berlin†	Schiff & Ziegler[3]			230(c)	97	95	27	11	26·56	8·63	64·81	-0·19
GERMANY	Stark & Henner 1953			1387	499	578	238	72	27·24	11·94	60·82	+2·43
GERMANY	Bonné-Tamir 1975		1.2	93	36	31	19	7	23·10	15·09	61·81	
GERMANY and SWITZERLAND	Levene 1975		1.2	514	186	202	92	34	26·53	13·15	60·32	
GREECE	Stark & Henner 1953	Sephardi.		144	32	62	31	19	33·82	19·18	47·00	-0·10
HUNGARY†	Darányi 1940	Includes half-Jews		342(c)	112	149	54	27	30·30	12·63	57·07	-0·21
HUNGARY	Stark & Henner 1953			141	57	55	22	7	25·20	10·89	63·91	+0·33
HUNGARY	Levene 1975		1.2	273	106	108	36	23	27·73	11·37	60·90	
IRISH REPUBLIC, Dublin†	Hooper 1947			78(c)	17	43	15	3	36·87	12·61	50·52	+2·18
NETHERLANDS†	Herwerden[3]			705	300	277	96	32	25·08	9·54	65·38	+0·36
NETHERLANDS†	Herwerden & Boele-Nijland 1930			142	63	58	12	9	27·10	7·63	65·27	-1·56
NETHERLANDS, Amsterdam†	Herzberger 1930[4]			1077(c)	456	435	126	60	26·41	9·01	64·58	-1·48
POLAND†	Halber & Mydlarski 1925			818(c)	271	339	142	66	28·92	13·64	57·44	-0·24
POLAND	Stark & Henner 1953			4633	1605	1962	771	295	28·46	12·29	59·25	+2·00
POLAND	Levene 1975		1.2	1310	443	537	225	105	28·56	13·49	57·95	
ROMANIA, Bucureşti†	Manuila, A. et al. 1945			2266	607	1055	440	164	32·31	14·48	53·21	+4·47
ROMANIA, Iaşi†	Ionescu & Ionescu 1930			1135	434	443	198	60	25·47	12·14	62·39	+1·49
ROMANIA, Maramureş†	Manuila, S. 1924			211(c)	55	82	42	32	31·77	19·16	49·16	-1·66
ROMANIA	Stark & Henner 1953			329	107	148	60	14	29·07	12·10	58·83	+2·43
ROMANIA	Levene 1975		1.2	1233	436	539	174	84	29·60	11·06	59·34	
UNITED KINGDOM, GREAT BRITAIN†	Roberts 1957-8	Mental patients		240	92	92	45	11	24·60	12·52	62·88	+1·21
YUGOSLAVIA, Monastir†	Hirszfeld & Hirszfeld 1919	Sephardi		500	194	165	116	25	21·39	15·36	63·25	+1·68
YUGOSLAVIA	Levene 1975		1.2	105	30	45	25	5	28·14	15·76	56·10	
U.S.S.R.												
BELORUSSIYA, Gomel†	Pevzner[5]			297(c)	96	125	50	26	29·82	13·71	56·47	-0·45
BELORUSSIYA, Minsk†	Rakovsky & Sukhotin[6]			257(c)	89	116	42	10	28·87	10·79	60·34	+1·90

[1] Quoted by Steffan & Wellisch 1936.
[2] Quoted by Boyd 1939.
[3] Quoted by Steffan & Wellisch 1932.
[4] Quoted by Bijlmer 1943.
[5] Quoted by Rubashkin 1929.
[6] Quoted by Semenskaya 1930.

64

TABLE 1.1 ABO SYSTEM: TESTS WITH ANTI-A AND -B (cont.)

Place	Population	Authors		Number	O	A	B	AB	p	q	r	D/σ
U.S.S.R. (cont.)												
BELORUSSIYA, Minsk†	Mental patients	Raskina 1930		94	33	41	16	4	28·03	11·36	60·61	+1·03
BELORUSSIYA, Mogilev†		Bunak[1]		116(c)	33	44	30	9	26·54	18·70	54·76	+0·96
LITHUANIA		Stark & Henner 1953		535	201	204	96	34	25·52	13·00	61·48	+0·33
LITHUANIA, Trakai†	Karaites	Reicher 1932		180	87	18	60	15	9·50	23·30	67·20	-2·96
LITHUANIA, Vilnius†	Karaites	Reicher 1932		108	46	20	36	6	12·87	21·83	65·30	+0·03
LITHUANIA, Trakai, Vilnius†	Karaites	Pulyanos 1963		51	13	10	23	5	16·04	32·96	51·00	+0·23
RUSSIA, Leningrad†		Smirnova & Chernyaeva 1929[3]		104	34	42	20	8	27·99	14·55	57·46	+0·20
RUSSIA, Moscow†		Grigorova 1931		371(c)	127	137	84	23	24·74	15·74	59·52	+1·39
RUSSIA, Moscow, Keshan†		Avdeyeva & Grytsevich[3]		350(c)	116	147	69	18	27·55	13·44	59·01	+1·98
RUSSIA		Stark & Henner 1953		1757	737	584	310	126	22·70	13·23	64·07	-2·41
RUSSIA		Levene 1975	1.2	576	201	231	97	47	27·97	13·36	58·67	
UKRAINE†		Lavrik et al. 1968		140(c)	61	57	17	5	25·42	8·21	66·37	+0·42
UKRAINE, Dnepropetrovsk†		Sakharov 1930[4]		114	27	51	27	9	31·63	17·53	50·84	+1·39
UKRAINE, Kharkov†		Eisenberg 1928		114	31	51	25	7	30·33	15·40	54·27	+1·48
UKRAINE, Kharkov†		Feldman & Elmanovich[3]		108(c)	33	36	19	20	29·70	19·48	50·82	-2·74
UKRAINE, Kharkov†		Gekker & Korochkan[5]		243(c)	91	101	39	12	26·97	11·16	61·87	+0·85
UKRAINE, Kharkov†		Rubashkin & Derman[3]		383(c)	107	162	90	24	28·66	16·41	54·93	+2·63
UKRAINE, Kherson†		Shiryak 1929		322(c)	113	130	66	13	25·72	13·27	61·01	+2·42
UKRAINE, Krym†	Karaites	Sabolotny 1928		373	136	113	96	28	21·15	18·32	60·53	+0·21
UKRAINE, Krym†	Krimchaks	Sabolotny 1928		500	104	172	163	61	27·16	25·93	46·91	+1·55
UKRAINE, Odessa†		Barinstein 1928		1475(c)	540	615	230	90	27·78	11·52	60·70	+0·56
UKRAINE, Odessa†		Leichik 1928		529	168	262	68	31	33·31	9·87	56·82	+0·82
DAGESTAN		Younovitch 1933b		87	35	23	21	8	19·62	18·21	62·17	-1·23
GEORGIA, Tbilisi	Georgian	Semenskaya et al. 1937b[6]		1983(a)	528	828	430	197	30·58	17·36	52·06	+1·24
UZBEKISTAN, Bukhara†		Grubina 1930		153	31	56	40	26	31·64	24·41	43·95	-0·67
UZBEKISTAN, Bukhara†		Kevorkov Martiukov 1940[7]		1000(c)	312	331	264	93	24·14	19·83	56·03	+0·37
UZBEKISTAN, Bukhara		Younovitch 1933b		121	36	37	39	9	21·47	22·54	55·99	+1·03
UZBEKISTAN, Samarkand†		Vishnevsky[8]		616(c)	199	180	188	49	20·82	21·65	57·53	+1·13
SIBERIA, Irkutsk†		Melkikh & Gringot 1926[9]		217(c)	69	89	41	18	28·82	14·68	56·50	+1·11
TURKEY												
TURKEY	Sephardi	Stark & Henner 1953		81	32	31	13	5	25·45	11·80	62·75	-0·07
TURKEY		Levene 1975	1.2	297	87	121	59	30	29·83	16·28	53·89	
TURKEY, Urfa	Kurdish	Horowitz 1963		92	12	53	20	7	42·44	16·50	41·06	+2·46
GREAT BRITAIN, CANADA, U.S.A., SOUTH AFRICA												
		Bar-Shany 1974		1899	709	781	284	125	27·66	11·40	60·94	-0·59
AMERICA												
CANADA, Manitoba†		Chown et al. 1949		140	48	62	24	6	28·53	11·45	60·02	+1·33
CANADA, Montreal†		Lubinski et al. 1945		967	339	406	135	87	29·77	12·14	58·09	-2·59

U.S.A., Brooklyn†	Wiener et al. 1929[10]		500	187	200	84	29	26·44	12·05	61·51	+0·63
U.S.A., New York City†	Schiff 1940	Mostly Jewish	117	43	48	20	6	26·75	11·87	61·38	+0·65
U.S.A., New York†	MacMahon & Folusiak 1958		375	142	165	46	22	29·15	9·51	61·34	−0·33
U.S.A., Ohio†	Rife 1957		523	202	216	77	28	27·01	10·61	62·38	+0·43
U.S.A.	Stark & Henner 1953		349	144	135	53	17	24·91	10·61	64·48	+0·41
ARGENTINA, Buenos Aires†	Neuman et al. 1961	Ashkenazi	8259	3283	3162	1305	509	25·45	11·65	62·90	−1·06
BRAZIL, São Paulo†	Ottensooser et al. 1963		100	35	37	21	7	25·23	15·19	59·58	+0·30
LATIN AMERICA	Bar-Shany 1974		1417	504	589	219	105	28·52	12·15	59·33	−0·87

(a), (c), †. For explanation of these symbols, used in this and subsequent tables, see Introduction to Tables (p. 59).

[1] Quoted by Wagner 1926.
[2] Quoted by Steffan & Wellisch 1932.
[3] Quoted by Hirszfeld 1928.
[4] Quoted by Steffan & Wellisch 1933.
[5] Quoted by Semenskaya 1930.
[6] Quoted by Boyd & Boyd 1937.
[7] Quoted by Gurevitch 1940.
[8] Quoted by Petrov 1928.
[9] Quoted by Jettmar 1930.
[10] Quoted by Steffan & Wellisch 1932.

TABLE 1.2
ABO SYSTEM: TESTS WITH ANTI-A, -A₁, AND -B

Place	Population	Authors	Number	O	A_1	A_2	B	A_1B	A_2B	p_1	p_2	q	r	χ^2
SOUTH-WEST ASIA														
ISRAEL	Samaritans	Levene 1975	1403	470	416	91	312	93	21	20·18	5·19	16·57	58·06	0·55
ISRAEL, Jaffa	Samaritans	Bonné 1966	132	89	9	18	15	1	0	3·87	7·39	6·27	82·47	1·57
ISRAEL, Nablus, Jaffa	Samaritans	Bonné-Tamir 1975	124*	66	18	25	14	1	0	7·98	11·78	6·28	73·96	2·38
ISRAEL, Nablus	Samaritans, Levi	Ikin et al. 1963	33(a)	8	12	8	4	1	0	22·20	17·94	8·01	51·85	1·51
ISRAEL, Nablus	Samaritans, Ephraim (44)	Ikin et al. 1963	49(a)	28	10	4	6	1	0	11·93	4·76	7·43	75·88	0·41
LEBANON and SYRIA		Levene 1975	204	72	68	9	41	10	4	21·53	4·17	14·57	59·73	1·87
IRAQ		Levene 1975	1147	256	414	35	305	127	10	27·42	2·80	21·72	48·06	3·04
IRAQ		Bonné-Tamir 1976	184	42	74	7	37	19	5	29·74	4·79	18·20	47·27	1·45
IRAQ, Hit	Karaites	Goldschmidt et al. 1976	72	10	4	18	34	2	4	4·26	18·27	34·58	42·89	5·55
IRAQ, S.E.	Kurdish	Tills et al. 1977	50	20	8	0	19	2	1	10·61	1·12	25·17	63·10	2·78
IRAN		Levene et al. 1977	154	51	38	15	37	10	3	17·01	7·39	17·84	57·76	0·41
IRAN		Levene et al. 1977	159	51	37	7	51	9	4	15·79	4·25	22·79	57·17	1·11
YEMEN		Dreyfuss et al. 1952	104	46	33	12	10	3	0	19·11	7·49	6·47	66·92	1·21
YEMEN, N. and N.E.		Tills et al. 1977	65	38	13	10	4	0	0	10·58	9·13	3·14	77·15	1·02
YEMEN, S.		Tills et al. 1977	92	57	7	13	10	2	3	5·00	9·50	8·43	77·07	4·48
YEMEN		Levene 1975	374	210	76	47	32	6	3	11·63	7·91	5·64	74·82	0·32
ADEN		Levene 1975	127	50	29	17	23	4	4	13·99	10·22	13·06	62·73	0·23
INDIA														
BOMBAY†	Bene-Israel	Sirsat 1956	200(c)	78	48	2	59	11	2	16·09	1·21	20·03	62·67	1·82
BOMBAY†	Baghdadi	Sirsat 1956	200(c)	45	55	5	68	21	6	21·38	3·60	27·58	47·44	2·03
AFRICA														
ALGERIA, Oran†		Auzas 1957	1445	580	391	60	310	90	14	18·29	3·18	15·50	63·03	1·16
EGYPT		Levene 1975	536	180	156	31	121	33	15	19·59	5·49	17·23	57·69	3·28
EGYPT	Karaites	Goldschmidt 1967	250							3·5	3·7	34·5	57·7	
LIBYA		Levene 1975	100	28	40	2	21	8	1	28·03	2·12	16·40	53·45	0·39
LIBYA		Bonné-Tamir 1974	148	44	68	7	18	11	0	31·62	3·55	10·32	54·51	1·49
MOROCCO		Levene 1975	426	159	99	9	128	24	7	15·72	2·25	20·85	61·18	3·66
MOROCCO, Marrakech†		Bonné-Tamir 1976	191	71	40	6	53	19	2	16·69	2·50	21·50	59·31	3·12
MOROCCO, Mogador†		Mechali et al. 1957	192	77	52	8	39	11	5	18·04	4·21	15·42	62·33	3·32
MOROCCO, Tafilalet†		Mechali et al. 1957	64	22	15	3	19	4	1	16·20	3·83	20·97	59·00	0·04
MOROCCO, Tafilalet†		Ikin et al. 1972	146	48	22	3	56	15	2	13·44	1·98	28·96	55·62	2·02
MOROCCO, Tafilalet. Rich†		Mechali et al. 1957	150	47	23	7	61	9	3	11·34	3·86	28·39	56·41	0·12
TUNISIA		Levene 1975	99	35	28	8	18	8	2	20·08	6·48	15·19	58·25	0·96
ETHIOPIA†	Falasha	Bat-Miriam 1962	152	65	45	13	22	5	2	18·10	6·27	10·06	65·57	0·04

TABLE 1.2 **ABO SYSTEM: TESTS WITH ANTI-A, -A$_1$, AND -B** (*cont.*)

Place	Population	Authors	Number	O	A$_1$	A$_2$	B	A$_1$B	A$_2$B	p$_1$	p$_2$	q	r	χ^2
EUROPE														
	AUSTRIA	Levene 1975	187	59	75	21	22	7	3	25·15	9·19	8·99	56·67	0·39
	BULGARIA	Levene 1975	283	72	116	15	52	23	5	28·77	5·16	15·35	50·72	0·30
	CZECHOSLOVAKIA	Levene 1975	380	161	110	30	56	18	5	18·52	5·83	10·97	64·68	0·61
	GERMANY, SWITZERLAND	Levene 1975	514	186	161	41	92	26	8	20·28	6·25	13·15	60·32	0·13
	GERMANY	Bonné-Tamir 1975	93	36	26	5	19	5	2	18·36	4·74	15·09	61·81	0·43
	HUNGARY	Levene 1975	273	106	87	21	36	15	8	20·80	6·93	11·37	60·90	4·89
	POLAND	Levene 1975	1310	443	446	91	225	82	23	22·73	5·83	13·49	57·95	0·41
	ROMANIA	Levene 1975	1233	436	423	116	174	69	15	22·42	7·18	11·06	59·34	2·47
	YUGOSLAVIA	Levene 1975	105	30	38	7	25	4	1	22·85	5·29	15·76	56·10	3·27
RUSSIA		Levene 1975	576	201	187	44	97	37	10	21·77	6·20	13·36	58·67	0·59
TURKEY		Levene 1975	297	87	101	20	59	24	6	23·87	5·96	16·28	53·89	0·08

*Data overlap with Ikin *et al.* 1963 and with Bonné 1966.

TABLE 1.3
ABO SYSTEM: TESTS WITH ANTI-A, -B, -A₁, AND -H; Aᵢ PRESENT

Place	IRAQ North-West	IRAN	YEMEN	
			Habban	Saada, San'a, Damar, Beida, Aden
Population	Kurdish	Kurdish		
Authors	Tills *et al.* 1977		Bonné *et al.* 1970	Bodmer *et al.* 1972
Number	61	106	595	202(a)
O	16	30	249	108
A_1	21	28	36	37
A_i	2	1	9	1
A_2	4	6	73	27
B	12	29	181	22
A_1B	4	10	12	5
A_iB	0	0	0	0
A_2B	2	2	35	2
p_1	23·26	19·84	4·12	10·99
p_i	2·19	0·58	0·79	0·27
p_2	6·97	4·89	10·02	8·48
q	16·08	21·66	21·34	7·45
r	51·50	53·03	63·73	72·81
χ^2	1·01	0·52	8·01	1·13

TABLE 2.1
MN SYSTEM: TESTS WITH ANTI-M AND -N

Place	Population	Authors		Number	MM	MN	NN	M	N	χ^2
SOUTH-WEST ASIA										
ISRAEL		Levene 1975	2.2	1398	510	617	271	58·55	41·45	
ISRAEL	Ashkenazi	Margolis *et al.* 1960		465	126	246	93	53·55	46·45	1·88
ISRAEL	Sephardi	Margolis *et al.* 1960		200	30	140	30	50·00	50·00	32·00
ISRAEL, Jaffa	Samaritans	Bonné 1966	2.2	131	15	75	41	40·08	59·92	
ISRAEL, Jaffa, Nablus	Samaritans	Bonné-Tamir 1975	2.3	124*	17	78	29	45·16	54·84	
ISRAEL, Nablus	Samaritans, Levi	Ikin *et al.* 1963	2.3	33(a)	5	17	11	40·91	59·09	
ISRAEL, Nablus	Samaritans, Ephraim (44)	Ikin *et al.* 1963	2.3	49(a)	9	25	15	43·88	56·12	
LEBANON, SYRIA		Levene 1975	2.2	203	63	96	44	54·68	45·32	
IRAQ		Levene 1975	2.2	1142	387	513	242	56·35	43·65	
IRAQ, Baghdad†		Boyd & Boyd 1941		215(c)	57	101	57	50·00	50·00	0·75
IRAQ, Baghdad		Gurevitch & Margolis 1955		162	66	64	32	60·49	39·51	4·88
IRAQ, Hit	Karaites	Goldschmidt *et al.* 1976	2.3	72	70	2	0	98·10	1·39	
IRAQ	Kurdish	Gurevitch & Margolis 1955		120	37	53	30	52·92	47·08	1·55
IRAQ, N.W.	Kurdish	Tills *et al.* 1977	2.2	61	29	20	12	63·93	36·07	
IRAQ, S.E.	Kurdish	Tills *et al.* 1977	2.2	50	11	21	18	43·00	57·00	
IRAN	Kurdish	Tills *et al.* 1977	2.2	106	30	60	16	56·60	43·40	
IRAN		Gurevitch *et al.* 1956		200(c)	63	111	26	59·25	40·75	4·46
IRAN		Levene 1975	2.2	152	50	73	29	56·91	43·09	
YEMEN		Brzezinski 1952		500	285	186	29	75·60	24·40	0·03
YEMEN		Dreyfuss *et al.* 1952	2.2	104	57	43	4	75·49	24·51	
YEMEN		Levene 1975	2.2	374	237	117	20	79·01	20·99	
YEMEN, N. & N.E.		Tills *et al.* 1977	2.2	65	47	16	2	84·62	15·38	

TABLE 2.1 MN SYSTEM: TESTS WITH ANTI-M AND -N (*cont.*)

Place	Population	Authors		Number	MM	MN	NN	M	N	χ^2
SOUTH-WEST ASIA (*cont.*)										
YEMEN, S.		Tills *et al.* 1977	2.2	92	60	25	7	78·81	21·19	
YEMEN, Habban		Bonné *et al.* 1970	2.2	553	345	178	30	78·48	21·52	
ADEN		Levene 1975	2.2	127	69	47	11	72·83	27·17	
INDIA										
KERALA, Cochin		Gurevitch *et al.* 1954		275	95	140	40	60·00	40·00	1·01
AFRICA										
EGYPT		Levene 1975	2.2	534	183	256	95	58·24	41·76	
LIBYA		Levene 1975	2.2	99	41	46	12	64·65	35·35	
LIBYA		Bonné-Tamir 1975	2.3	148	48	74	26	57·43	42·57	
LIBYA, Tripolitania		Gurevitch *et al.* 1954		200	55	92	53	50·50	49·50	1·28
MOROCCO		Margolis *et al.* 1957		220(c)	53	140	27	55·91	44·09	18·60
MOROCCO		Levene 1975	2.2	423	171	192	60	63·12	36·88	
MOROCCO		Bonné-Tamir 1976	2.3	191	81	80	30	63·35	36·65	
MOROCCO, Tafilalet†		Ikin *et al.* 1972	2.2	146	68	65	13	68·88	31·12	
TUNISIA	'Tonensa'	Margolis *et al.* 1957		200(c)	58	106	36	55·50	44·50	1·06
TUNISIA		Levene 1975	2.2	98	33	48	17	58·16	41·84	
TUNISIA, Île de Djerba†		Moullec & Abdelmoula 1954		70	32	33	5	69·29	30·71	0·81
ETHIOPIA†	Falasha	Bat-Miriam 1962		152	69	62	21	65·79	34·21	1·34
EUROPE										
AUSTRIA		Levene 1975	2.2	187	64	89	34	58·02	41·98	
BULGARIA		Levene 1975	2.2	283	83	146	54	55·12	44·88	
CZECHOSLOVAKIA		Levene 1975	2.2	377	142	177	58	61·14	38·86	
GERMANY		Levene 1975	2.2	513	151	252	110	54·00	46·00	
GERMANY		Bonné-Tamir 1975	2.3	92	28	49	15	57·06	42·94	
HUNGARY		Levene 1975	2.2	271	85	128	58	54·98	45·02	
POLAND		Levene 1975	2.2	1305	402	642	261	55·40	44·60	
ROMANIA		Levene 1975	2.2	1221	402	593	226	57·20	42·80	
YUGOSLAVIA		Levene 1975	2.2	105	33	51	21	55·71	44·29	
U.S.S.R.										
LITHUANIA, Trakai, Vilnius†	Karaites	Pulyanos 1963		51	27	17	7	69·61	30·39	2·30
RUSSIA		Levene 1975	2.2	577	204	273	100	59·01	40·99	
UKRAINE†		Lavrik *et al.* 1968		21(c)	7	10	4	57·14	42·86	0·02
TURKEY		Levene 1975	2.2	297	91	155	51	56·73	43·27	
AMERICA										
U.S.A., Ohio†	Of Central European and Russian origin	Rife & Schonfeld 1944		84(c)	32	44	8	64·29	35·71	1·66
BRAZIL, Sâo Paulo†	Ashkenazi, mainly from Poland	Ottensooser *et al.* 1963		100	25	54	21	52·00	48·00	0·67

*Data overlap with Ikin *et al.* 1963 and with Bonné 1966.

TABLE 2.2
MN SYSTEM: TESTS WITH ANTI-M, -N, AND -S

Place	Population	Authors	Number	MMS	MsMs	MNS	MsNs	NNS	NsNs	MS	Ms	NS	Ns	χ^2
SOUTH-WEST ASIA														
ISRAEL		Levene 1975	1398	371	139	393	224	119	152	28·59	29·96	10·79	30·66	12·63
ISRAEL, Jaffa	Samaritans	Bonné 1966	131*	12	3	49	26	26	15	19·61	20·47	22·37	37·55	5·44
LEBANON and SYRIA		Levene 1975	203	48	15	56	40	18	26	26·95	27·73	9·99	35·33	0·78
IRAQ		Levene 1975	1142	288	99	358	155	128	114	28·85	27·50	14·43	29·22	11·19
IRAQ, N.W.	Kurdish	Tills et al. 1977	61	15	14	10	10	7	5	18·99	44·94	12·07	24·00	5·23
IRAQ, S.E.	Kurdish	Tills et al. 1977	50	10	1	14	7	9	9	26·41	16·59	15·76	41·24	1·68
IRAN	Kurdish	Tills et al. 1977	106	13	17	35	25	8	8	15·70	40·90	15·60	27·80	3·68
IRAN		Levene 1975	152	37	13	53	20	18	11	28·80	28·11	17·39	25·70	0·35
YEMEN		Dreyfuss et al. 1952	104	38	19	21	22	1	3	31·75	43·74	3·09	21·42	1·43
YEMEN		Levene 1975	374	170	67	76	41	10	10	37·30	41·71	6·54	14·45	1·36
YEMEN, N. & N.E.		Tills et al. 1977	65	28	19	6	10		2	30·98	53·64		15·38	0·21
YEMEN, S.		Tills et al. 1977	92	46	14	15	10	5	2	39·21	39·60	7·71	13·48	4·63
YEMEN, Habban		Bonné et al. 1970	553	233	112	79	99	3	27	33·47	45·01	1·06	20·46	1·34
ADEN		Levene 1975	127	49	20	36	11	4	7	36·47	36·36	8·78	18·39	4·93
AFRICA														
EGYPT		Levene 1975	534	128	55	150	106	42	53	26·22	32·02	10·47	31·29	0·10
LIBYA		Levene 1975	99	37	4	33	13	8	4	41·30	23·35	12·78	22·57	1·40
MOROCCO		Levene 1975	423	112	59	118	74	24	36	27·39	35·73	9·38	27·50	2·27
MOROCCO, Tafilalet Oases†		Ikin et al. 1972	146	43	25	31	34	6	7	25·99	42·89	6·83	24·29	0·94
TUNISIA		Levene 1975	98	22	11	33	15	9	8	26·36	31·80	14·72	27·12	0·74
EUROPE														
AUSTRIA		Levene 1975	187	45	19	48	41	17	17	24·79	33·23	11·09	30·89	1·18
BULGARIA		Levene 1975	283	59	24	98	48	27	27	26·74	28·38	14·07	30·81	1·33
CZECHOSLOVAKIA		Levene 1975	377	94	48	116	61	27	31	27·17	33·97	11·87	26·99	2·54
GERMANY		Levene 1975	513	99	52	141	111	39	71	23·01	30·99	9·45	36·55	0·56
HUNGARY		Levene 1975	271	66	19	74	54	19	39	28·05	26·93	7·85	37·17	0·88
POLAND		Levene 1975	1305	279	123	385	257	106	155	25·35	30·05	10·61	33·99	0·93
ROMANIA		Levene 1975	1221	273	129	350	243	104	122	24·92	32·28	11·48	31·32	0·13
YUGOSLAVIA		Levene 1975	105	24	9	34	17	7	14	29·10	26·61	9·05	35·24	0·97
U.S.S.R.														
RUSSIA		Levene 1975	577	147	57	161	112	54	46	26·71	32·30	12·27	28·72	1·87
TURKEY														
		Levene 1975	297	64	27	95	60	23	28	26·18	30·55	11·55	31·72	1·26

*Data overlap with Bonné-Tamir 1975.

TABLE 2.3
MN SYSTEM: TESTS WITH ANTI-M, -N, -S, AND -s

Place	ISRAEL				IRAQ	IRAN	YEMEN	LIBYA	MOROCCO	GERMANY
	Nablus		Jaffa, Nablus		Hit		Saada, San'a, Damar, Beida, Aden			
	(SOUTH-WEST ASIA)							*(NORTHERN AFRICA)*		*(EUROPE)*
Population	*Samaritans, Levi*	*Samaritans, Ephraim*	*Samaritans*		*Karaites*					
Authors	Ikin *et al.* 1963		Bonné-Tamir 1975		Goldschmidt *et al.* 1976	Levene *et al.* 1977	Bodmer *et al.* 1972		Bonné-Tamir 1976	
Number	33(a)	49(a)	124*	184	72	147	202(a)	148	191	92
MMSS	3	0	6	15	2	10	16	14	18	8
MMSs	2	6	6	28	22	22	53	21	38	9
MMss	0	3	5	17	46	11	50	13	25	11
MNSS	2	0	18	14		14	6	10	9	3
MNSs	8	11	34	46	0	26	25	32	26	27
MNss	7	14	26	31	2	21	38	32	45	19
NNSS	0	0	3	4		8	0	1	2	2
NNSs	0	3	13	15		21	7	9	9	2
NNss	11	12	13	14	0	14	7	16	19	11
MS	26·92	15·80	23·66	27·59	18·05	23·94	26·52	28·25	26·76	26·97
Ms	13·99	28·08	21·50	29·74	80·05	26·06	49·47	29·18	36·59	30·09
NS	3·38	4·61	19·49	14·53	1·39	21·30	5·41	9·59	7·54	7·81
Ns	55·71	51·51	35·35	28·14		28·70	18·60	32·98	29·11	35·13
χ^2	6·65	3·46	12·44	0·38	0·57	5·33	4·06	2·34	8·43	10·49

* Data overlap with Ikin *et al.* 1963 and with Bonné 1966.

TABLE 2.4
MN SYSTEM: TESTS WITH ANTI-M^g

Place	Population	Authors	Number	M^g+
ISRAEL, Jaffa	Samaritans	Bonné 1966	131	0

TABLE 2.5
MN SYSTEM: TESTS WITH ANTI-He (HENSHAW)

Place	Population	Authors	Number	He+	He+%
SOUTH-WEST ASIA					
ISRAEL, Nablus	Samaritans	Ikin *et al.* 1963	82	0	
IRAQ, N.W.	Kurdish	Tills *et al.* 1977	61	0	
IRAQ, S.E.	Kurdish	Tills *et al.* 1977	50	0	
IRAN	Kurdish	Tills *et al.* 1977	106	0	
YEMEN		Dreyfuss *et al.* 1952	104	0	
YEMEN, N. and N.E.		Tills *et al.* 1977	65	0	
YEMEN, S.		Tills *et al.* 1977	92	7	7·61
YEMEN, Habban		Bonné *et al.* 1970	553	0	
AFRICA					
TUNISIA, Île de Djerba†		Moullec & Abdelmoula 1954	70	0	

TABLE 2.6
MN SYSTEM: TESTS WITH ANTI-Mt^a (MARTIN)

Place	Population	Authors	Number	Mt(a+)
ISRAEL	Samaritans	Konugres *et al.* 1965	132	0

TABLE 3
P SYSTEM: TESTS WITH ANTI-P$_1$

Place	Population	Authors	Number	P$_1$	P$_2$	P$_1$	P$_2$+p
SOUTH-WEST ASIA							
ISRAEL, Jaffa	Samaritans	Bonné 1966	124	46	78	20·69	79·31
ISRAEL, Nablus	Samaritans, Levi	Ikin 1963	33(a)	12	21	20·23	79·77
ISRAEL, Nablus	Samaritans, Ephraim	Ikin 1963	49(a)	17	32	19·19	80·81
ISRAEL, Nablus, Jaffa	Samaritans	Bonné-Tamir 1975	124*	54	70	24·87	75·13
IRAQ		Bonné-Tamir 1976	184	141	43	51·66	48·34
IRAQ, N.W.	Kurdish	Tills et al. 1977	61	30	31	28·71	71·29
IRAQ, S.E.	Kurdish	Tills et al. 1977	50	19	31	21·26	78·74
IRAN	Kurdish	Tills et al. 1977	106	71	35	42·54	57·46
YEMEN, N. & N.E.		Tills et al. 1977	65	34	31	30·94	69·06
YEMEN, S.		Tills et al. 1977	92	55	37	36·58	63·42
YEMEN, Habban		Bonné et al. 1970	587	400	187	43·56	56·44
YEMEN, Saada, San'a, Damar, Beida, Aden		Bodmer et al. 1972	202(a)	134	68	41·98	58·02
NORTHERN AFRICA							
LIBYA		Bonné-Tamir 1974	148	108	40	48·01	51·99
MOROCCO		Bonné-Tamir 1976	191	144	47	50·39	49·61
MOROCCO, Tafilalet†		Ikin et al. 1972	146	70	76	27·85	72·15
EUROPE							
GERMANY		Bonné-Tamir 1974	93	69	24	49·20	50·80

*Data overlap with Ikin 1963 and with Bonné 1966.

TABLE 4.1
RHESUS SYSTEM: TESTS WITH ANTI-D ..

Place	Population	Authors		Number	D+	D−	$\bar{D}$−	D	d
SOUTH-WEST ASIA									
ISRAEL	Parents born there; sampled in Marseille	Bar-Shany 1974		4205	3820	385	9·16	69·73	30·27
ISRAEL		Lévy et al. 1967		40	39	1	2·50	84·19	15·81
ISRAEL	Born there	Levene 1975	4.8	1401	1290	111	7·92	71·86	28·14
ISRAEL	Ashkenazi	Gurevitch et al. 1947		1035	921	114	11·01	66·82	33·18
ISRAEL	Ashkenazi	Gurevitch et al. 1951	4.4	946(c)	807	139	14·69	61·67	38·33
ISRAEL	Ashkenazi	Margolis et al. 1960	4.5	465	421	44	9·46	69·24	30·76
ISRAEL	Sephardi	Gurevitch et al. 1947		262	231	31	11·83	65·61	34·39
ISRAEL	Sephardi	Gurevitch et al. 1951	4.4	252(c)	219	33	13·09	53·82	36·18
ISRAEL	Sephardi	Margolis et al. 1960	4.5	200	179	21	10·50	67·60	32·40
ISRAEL	Oriental	Gurevitch et al. 1947		369	335	34	9·21	69·65	30·35
ISRAEL	Oriental	Gurevitch et al. 1951	4.4	137(c)	118	19	13·87	62·76	37·24
ISRAEL, Nablus	Samaritans, Levi	Ikin 1963	4.8	33(a)	31	2	6·06	75·38	24·62
ISRAEL, Nablus	Samaritans, Ephraim	Ikin 1963	4.8	49(a)	40	9	18·37	57·14	42·86
ISRAEL, Jaffa	Samaritans	Bonné 1966	4.7	132	107	25	18·94	56·48	43·52
ISRAEL, Nablus, Jaffa	Samaritans	Bonné-Tamir 1975	4.7	124*	107	17	13·71	62·97	37·03
LEBANON and SYRIA									
		Levene 1975	4.7	203	183	20	9·85	68·62	31·38
IRAQ	Females	Silberstein & Goldstein 1958		308(c)	276	32	10·39	67·77	32·23
IRAQ		Levene 1975	4.8	1146	1063	83	7·24	73·09	26·91
IRAQ, Baghdad		Bonné-Tamir 1976	4.8	185	175	10	5·54	76·46	23·54
IRAQ, Hit		Gurevitch & Margolis 1954-5	4.5	162	149	13	8·02	71·68	28·32
	Karaites	Goldschmidt et al. 1976	4.7	72	52	20	27·78	47·29	52·71
IRAQ	Kurdish	Gurevitch et al. 1953	4.4	250	231	19	7·60	72·43	27·57
IRAQ	Kurdish	Gurevitch & Margolis 1954-5	4.5	129	124	5	3·88	80·30	19·70
IRAQ, N.W.	Kurdish	Tills et al. 1977	4.9	61	56	5	8·20	71·36	28·64
IRAQ, S.E.	Kurdish	Tills et al. 1977	4.9	50	49	1	2·00	85·86	14·14
IRAN	Kurdish	Tills et al. 1977	4.9	106	99	7	6·60	74·31	25·69
IRAN and IRAQ	Kurdish	Gurevitch et al. 1947		203	200	3	1·48	87·83	12·17
IRAN		Silberstein & Goldstein 1958		225	214	11	4·89	77·89	22·11
IRAN		Gurevitch et al. 1956	4.5	200	190	10	5·00	77·64	22·36
IRAN		Levene 1975	4.7	152	143	9	5·92	75·67	24·33
IRAN		Levene et al. 1977	4.7	159	150	9	5·66	76·21	23·79
IRAN and IRAQ		Bar-Shany 1974		4983	4633	350	7·02	73·50	26·50
YEMEN	Females	Silberstein & Goldstein 1958		167(c)	143	24	14·37	62·09	37·91
YEMEN		Brzezinski et al. 1952	4.5	500	450	50	10·00	68·38	31·62

TABLE 4.1 SYSTEM: TESTS WITH ANTI-D (cont.)

Place	Population	Authors		Number	D+	D−	D̄−	D	d
SOUTH-WEST ASIA (cont.)									
YEMEN		Bar-Shany 1974		2309	2064	235	10·18	68·09	31·91
YEMEN		Levene 1975		373	327	46	12·33	64·89	35·11
YEMEN		Dreyfuss et al. 1952	4.8	104	100	4	3·85	80·38	19·62
YEMEN		Gurevitch et al. 1947	4.8	44	43	1	2·27	84·93	15·07
YEMEN, N. and N.E.		Tills et al. 1977		65	62	3	4·62	78·51	21·49
YEMEN, S.		Tills et al. 1977		92	87	5	5·43	76·70	23·30
YEMEN, Habban		Bonné et al. 1970	4.10	596	559	37	6·21	75·08	24·92
ADEN		Levene 1975	4.7	127	117	10	7·87	71·95	28·05
INDIA									
BOMBAY†	Baghdadi	Bar-Shany 1974		404	338	66	16·34	59·58	40·42
BOMBAY†	Bene-Israel	Sirsat 1956		200(c)	181	19	9·50	69·18	30·82
KERALA, Cochin		Sirsat 1956		200(c)	177	23	11·50	66·09	33·91
		Gurevitch et al. 1954	4.5	275	221	54	19·64	55·68	44·32
AFRICA									
ALGERIA, Oran†	Autochthons or from Tetouan	Solal & Hanoun 1952	4.6	150(c)	122	28	18·67	56·79	43·21
ALGERIA, Oran†	Parents born there; sampled in Marseille	Auzas 1956		100	88	12	12·00	65·36	34·64
ALGERIA		Lévy et al. 1967		128	117	11	8·59	70·69	29·31
ALGERIA		Bar-Shany 1974		643	576	67	10·42	67·72	32·28
EGYPT		Levene 1975	4.8	534	489	45	8·43	70·97	29·03
LIBYA		Bar-Shany 1974		1071	959	112	10·46	67·66	32·34
LIBYA		Bonné-Tamir 1976	4.7	148	125	23	15·54	60·58	39·42
LIBYA		Levene 1975	4.7	99	84	15	15·15	61·08	38·92
LIBYA, Tripolitania	Females	Silberstein & Goldstein 1958		193(c)	168	25	12·95	64·01	35·99
LIBYA, Tripolitania		Gurevitch et al. 1954	4.5	200	171	29	14·50	61·92	38·08
MOROCCO	Females	Silberstein & Goldstein 1958		1014	895	119	11·74	65·74	34·26
MOROCCO	Parents born there; sampled in Marseille	Lévy et al. 1967		60	49	11	18·33	57·19	42·81
MOROCCO		Margolis et al. 1957	4.5	220(c)	196	24	10·91	66·97	33·03
MOROCCO		Bar-Shany 1974		6089	5369	720	11·82	65·62	34·38
MOROCCO		Levene 1975	4.7	421	378	43	10·21	68·05	31·95
MOROCCO		Bonné-Tamir 1976	4.7	192	179	13	6·77	73·98	26·02
MOROCCO, Marrakech†		Mechali et al. 1957	4.5	192	165	27	14·06	62·50	37·50
MOROCCO, Mogador†		Mechali et al. 1957	4.5	64	57	7	10·94	66·92	33·08
MOROCCO, Rabat†		Mechali et al. 1957		526	461	65	12·36	64·85	35·15
MOROCCO, Tafilalet†		Ikin et al. 1972	4.10	146	133	13	8·90	70·17	29·83
MOROCCO, Tafilalet, Erfoud†		Mechali et al. 1957	4.7	200	176	24	12·00	65·36	34·64
MOROCCO, Tafilalet, Gourrama†		Mechali et al. 1957		136	121	15	11·03	66·79	33·21
MOROCCO, Tafilalet, Rich†		Mechali et al. 1957	4.7	150	134	16	10·67	67·34	32·66
TUNISIA	Parents born there; sampled in Marseille	Silberstein & Goldstein 1958		408(c)	353	55	13·48	63·28	36·72
TUNISIA		Lévy et al. 1967		49	39	10	20·41	54·82	45·18

*Data overlap with Ikin 1963 and with Bonné 1966.

TABLE 4.1 RHESUS SYSTEM: TESTS WITH ANTI-D (*cont.*)

Place	Population	Authors		Number	D+	D−	D̄−	D	d
AFRICA (*cont.*)									
TUNISIA		Margolis et al. 1957	4.5	200(c)	181	19	9·50	69·18	30·82
TUNISIA		Bar-Shany 1974		1584	1421	163	10·29	67·92	32·08
TUNISIA		Levene 1975	4.7	98	85	13	13·27	63·57	36·43
TUNISIA, Île de Djerba†		Moullec & Abdelmoula 1954	4.5	70	58	12	17·14	58·60	41·40
TUNISIA, Île de Djerba, Hara-Kbira†		Ranque et al. 1964		111	97	14	12·61	64·49	35·51
TUNISIA, Île de Djerba, Hara-Srira†		Ranque et al. 1964		60	53	7	11·67	65·84	34·16
ETHIOPIA†	Falasha	Bat-Miriam 1962	4.8	152	143	9	5·92	75·67	24·33
EUROPE									
WESTERN		Silberstein & Goldstein 1958		222(c)	198	24	10·81	67·12	32·88
SOUTHERN and SOUTH-EASTERN		Bar-Shany 1974		4809	4329	480	9·98	68·41	31·59
EASTERN		Bar-Shany 1974		2598	2321	277	10·66	67·35	32·65
		Bar-Shany 1974		25 637	23 348	2289	8·93	70·12	29·88
AUSTRIA		Levene 1975	4.7	187	175	12	6·42	74·66	25·34
BULGARIA		Levene 1975	4.7	283	265	18	6·36	74·78	25·22
CZECHOSLOVAKIA		Levene 1975	4.8	375	349	26	6·93	73·68	26·32
GERMANY		Levene 1975	4.7	513	465	48	9·36	69·41	30·59
GERMANY		Bonné-Tamir 1976	4.7	93	80	13	13·98	62·61	37·39
HUNGARY		Levene 1975	4.8	272	253	19	6·99	73·56	26·44
POLAND		Levene 1975	4.8	1306	1188	118	9·04	69·93	30·07
ROMANIA		Levene 1975	4.8	1226	1133	93	7·59	72·45	27·55
UNITED KINGDOM, GREAT BRITAIN†	Mental patients	Roberts 1957-8		228	210	18	7·89	71·90	28·10
YUGOSLAVIA		Levene 1975	4.7	105	94	11	10·48	67·63	32·37
U.S.S.R.									
LITHUANIA, Trakai, Vilnius†	Karaites	Levene 1975	4.7	577	525	52	9·01	69·98	30·02
		Pulyanos 1963		51(c)	44	7	13·73	62·95	37·05
TURKEY									
TURKEY, Urfa	Kurdish	Levene 1975	4.8	298	277	21	7·05	73·45	26·55
		Horowitz 1963		92	80	12	13·04	63·89	36·11
GREAT BRITAIN, CANADA, U.S.A., SOUTH AFRICA		Bar-Shany 1974		1899	1698	201	10·58	67·47	32·53
NORTH AMERICA									
CANADA, Montreal†	Antenatal	Lubinski et al. 1945		514	472	42	8·17	71·42	28·58
CANADA, Manitoba†	Of Russian or Central European origin	Chown et al. 1949	4.5	140	127	13	9·29	69·52	30·48
U.S.A. Ohio†		Rife & Schonfeld 1944		40(c)	36	4	10·00	68·38	31·62
LATIN AMERICA									
BRAZIL, São Paulo†		Bar-Shany 1974		1417	1313	104	7·34	72·91	27·09
	Ashkenazi	Ottensooser et al. 1962	4.5	100	96	4	4·00	80·00	20·00

TABLE 4.2
RHESUS SYSTEM: TESTS WITH ANTI-Cᵂ; Cᵂ+ABSENT

Place	Population	Authors	Number
YEMEN		Dreyfuss *et al.* 1952	104
YEMEN, Habban		Bonné *et al.* 1970	559

TABLE 4.3
RHESUS SYSTEM: TESTS WITH ANTI-V; V+ ABSENT

Place	Population	Authors	Number
ISRAEL, Nablus	Samaritans	Ikin *et al.* 1963	79(a)

TABLE 4.4
RHESUS SYSTEM: TESTS WITH ANTI-C, -D, AND -E

Place	ISRAEL					IRAQ	
Population	*Ashkenazi*		*Sephardi*		*Oriental*	*Kurdish*	
Authors	Gurevitch *et al.* 1951					Gurevitch *et al.* 1953	
Number	946 (c)		252 (c)		137 (c)	250	
	Obs.	Exp.	Obs.	Exp.	Obs.	Exp.	Obs.	Exp.

	Obs.	Exp.	Obs.	Exp.	Obs.	Exp.	Obs.	Exp.
CDE	123	112·11	19	16·53	10	12·11	54	45·95
CDee	546	554·07	169	170·86	88	86·52	103	109·00
CddE								0·27
Cddee	11	11·16	2	2·02	1	0·99	2	1·78
ccDE	103	112·67	13	15·35	14	12·08	48	54·07
ccDee	35	33·58	18	17·38	6	6·34	26	23·03
ccddE							4	4·15
ccddcc	128	122·41	31	29·86	18	18·96	13	11·75
CDe		45·11		49·02		46·81		37·42
Cde		1·60		1·14		0·95		1·59
cDE		12·69		6·54		9·26		20·15
cDe		4·64		8·87		5·77		15·61
cdE								3·55
cde		35·97		34·43		37·21		21·68

TABLE 4.5
RHESUS SYSTEM: TESTS WITH ANTI-C, -D, -E, AND -c

	SOUTH-WEST ASIA												INDIA	
Place	ISRAEL				IRAQ Baghdad		Kurdish		IRAN		YEMEN		KERALA Cochin	
Population	Ashkenazi		Sephardi				Kurdish							
Authors	Margolis et al. 1960				Gurevitch & Margolis 1954–5				Gurevitch et al. 1956		Brzezinski et al. 1952		Gurevitch et al. 1954	
Number	465		200		162		129		200(c)		500		275	
	Obs.	Exp.	Obs.	Exp.	Obs.	Exp.	Obs.	Exp.	Obs.	Exp.	Obs.	Exp.	Obs.	Exp.
CCDE	2	2·09	2	1·88	9	8·75	14	14·49			3	3·05		
CCDee	125	128·66	54	52·32	55	47·53	38	35·20	73	73·24	173	162·40	45	54·09
CCddee				0·44		0·05						0·05		0·16
CcDE	59	55·99	21	22·64	34	35·67	34	32·53	26	26·28	47	46·85	12	12·32
CcDee	179	174·70	70	71·90	28	41·44	22	28·21	70	69·32	173	194·20	135	117·59
CcddE				0·14		0·08								
Ccddee	1	0·98	6	5·10	1	1·20					3	3·30	7	5·89
ccDE	40	42·64	18	16·64	18	16·85	12	12·91	15	14·80	30	30·20	15	14·71
ccDee	16	16·23	14	13·76	5	2·97	4	2·50	6	6·12	24	20·20	14	16·09
ccddE			1	0·80	1	0·87								
ccddee	43	43·71	14	14·32	11	6·59	5	3·15	10	10·26	47	39·75	47	54·15
CDE	0·43		0·91		4·77		9·82				0·53			
CDe	52·26		46·61		52·37		52·24		60·52		55·85		42·01	
Cde	0·34		4·76		1·83						1·15		2·41	
cDE	11·06		9·50		15·46		17·01		10·86		7·83		5·04	
cDe	5·25		10·71		4·11		5·32		5·97		6·44		6·17	
cdE			0·74		1·29									
cde	30·66		26·76		20·17		15·61		22·65		28·20		44·37	

TABLE 4.5 **RHESUS SYSTEM: TESTS WITH ANTI-C, -D, -E, AND -c** (*cont.*)

	AFRICA											
Place	**LIBYA** **Tripolitania**				**Marrakech†**		**MOROCCO** **Mogador†**		**Tafilalet Oases** **Erfoud†**		**Rich†**	
Population												
Authors	Gurevitch *et al.* 1954		Margolis *et al.* 1957		Mechali *et al.* 1957							
Number	200		220(c)		192		64		200		150	
	Obs.	Exp.	Obs.	Exp.	Obs.	Exp.	Obs.	Exp.	Obs.	Exp.	Obs.	Exp.
CCDE	6	6·96			3	2·88	3	2·04	3	2·94	1	1·23
CCDee	38	35·94	68	62·88	41	43·12	15	12·05	50	51·92	43	46·32
CCddee					1	0·08				0·34		0·06
CcDE	23	19·52	15	14·94	17	17·68	4	6·91	22	22·56	12	9·42
CcDee	74	78·92	84	94·53	80	75·59	20	24·97	76	72·08	67	62·60
CcddE												
Ccddee					1	2·69			6	5·10	2	1·79
ccDE	13	14·44	12	12·12	15	14·50	7	5·10	17	16·56	4	6·23
ccDee	17	16·00	17	14·74	9	9·64	8	6·90	8	8·66	7	7·43
ccddE												
ccddee	29	27·22	24	20·79	25	25·82	7	6·03	18	19·84	14	14·91
CDE		3·92				1·56		3·53		1·42		0·76
CDe		42·39		53·46		45·52		43·40		47·06		53·72
Cde						1·91				4·05		1·88
cDE		7·21		6·35		8·03		8·13		9·72		5·05
cDe		9·59		9·44		6·31		14·25		6·25		7·06
cdE												
cde		36·89		30·75		36·67		30·69		31·50		31·53

TABLE 4.5 RHESUS SYSTEM: TESTS WITH ANTI-C, -D, -E, AND -c (*cont.*)

	AFRICA (*cont.*)				AMERICA			
Place	TUNISIA		Île de Djerba†		CANADA Manitoba†		BRAZIL São Paulo†	
Population							*Ashkenazi*	
Authors	Margolis *et al.* 1957		Moullec & Abdelmoula 1954		Chown *et al.* 1949		Ottensooser *et al.* 1962	
Number	200(c)		70		140		100	
	Obs.	Exp.	Obs.	Exp.	Obs.	Exp.	Obs.	Exp.
CCDE	1	0·96			1	1·32	1	0·98
CCDee	68	62·92	19	16·92	41	43·79	25	24·55
CCddee				0·12		0·08		
CcDE	15	15·50	7	7·32	24	18·40	17	17·43
CcDee	73	82·80	22	25·64	49	48·47	32	32·57
CcddE								
Ccddee			2	2·03	2	1·82		
ccDE	11	10·58	7	6·73	7	11·70	14	13·67
ccDee	13	11·06	3	2·57	5	4·48	7	6·86
ccddE								
ccddee	19	16·18	10	8·67	11	9·94	4	3·94
CDE		0·43				0·83		0·98
CDe		56·09		45·21		53·53		49·55
Cde				4·12		2·45		
cDE		6·58		10·60		11·10		16·60
cDe		8·46		4·88		5·44		13·02
cdE								
cde		28·44		35·19		26·65		19·85

TABLE 4.6
RHESUS SYSTEM: TESTS WITH ANTI-C, -D, -E, AND -c AND FOR D^u

AFRICA		
Place	ALGERIA **Oran†**	
Authors	Auzas 1956	
Number	100	
	Obs.	Exp.
CCDE	8	5·28
CCDee	16	16·35
CcDE	9	16·31
CcDee	36	33·44
ccDE	14	10·07
ccD^{u}E		0·16
ccDee	2	2·28
ccD^{u}ee	3	3·18
ccddE	2	1·28
ccddee	10	11·64
CDE	6·07	
CDe	40·44	
cDE	10·31	
cDe	2·86	
cD^{u}e	4·38	
cdE	1·83	
cde	34·11	

TABLE 4.7
RHESUS SYSTEM: TESTS WITH ANTI-C, -D, -E, -c, AND -e

	SOUTH-WEST ASIA													NORTHERN AFRICA			
Place	ISRAEL				LEBANON, SYRIA		IRAQ Hit		IRAN				ADEN		LIBYA			
	Jaffa		Jaffa, Nablus															
Population	Samaritans		Samaritans				Karaites											
Authors	Bonné 1966		Bonné-Tamir 1975		Levene 1975		Goldschmidt et al. 1976		Levene 1975		Levene et al. 1977b		Levene 1975		Levene 1975		Bonné-Tamir 1976	
Number	132		124*		203		72		152		159		127		99		148	
	Obs.	Exp.	Obs.	Exp.	Obs.	Exp.	Obs.	Exp.	Obs.	Exp.	Obs.	Exp.	Obs.	Exp.	Obs.	Exp.	Obs.	Exp.
CCDEE						0·02				0		0		0·06		0·01		
CCDEe					2	2·25			1	0·79	1	0·72	3	2·81	1	0·83		
CCDee	24	24·20	31	29·04	54	52·39	4	3·79	51	51·28	43	44·15	27	34·04	26	26·52	43	39·96
CCddee						0·12				0·03		0·03		0·05		0·01		0·10
CcDEE					1	0·55				0·23		0·30		0·42		0·03		0·10
CcDEe	12	9·85	9	11·12	26	27·22	1	1·15	31	29·75	41	37·05	14	12·24	2	3·02	4	7·28
CcDee	53	54·79	49	50·80	69	70·65	24	24·30	42	42·85	40	40·89	61	48·84	49	45·19	61	63·69
CcddEe						0·08												
Ccddee					3	2·84			1	0·96	1	0·95	2	1·61	1	1·01		
ccDEE		1·00		1·07	2	3·15		0·09	4	4·21	5	7·62	1	0·75		0·05	3	2·92
ccDEe	11	11·14	14	9·74	20	17·03	4	3·67	12	12·55	18	17·38	6	7·52	3	1·93		0·33
ccDee	7	6·78	4	4·24	9	9·28	19	19·01	2	1·87	3	2·96	5	7·26	3	3·44	10	6·05
ccddEE						0·01												
ccddEe					1	0·97											7	7·16
ccddee	25	24·24	17	17·99	16	16·44	20	19·99	8	7·48	7	6·95	8	11·40	14	15·95	20	20·51
CDE						1·09				0·45		0·43		2·13		0·80		
CDe		42·81		48·39		48·40		22·93		56·67		51·28		49·69		51·47		49·38
Cde						2·46				1·43		1·44		2·12		1·27		
cDE		8·71		9·27		11·61		3·47		16·65		21·89		7·72		2·21		2·65
cDe		5·63		4·25		7·14		20·90		2·62		4·06		8·37		4·11		4·73
cdE						0·85												6·01
cde		42·85		38·09		28·45		52·70		22·18		20·90		29·97		40·14		37·23
		No D[u]		No D[u]		No D[u]				No D[u]				No D[u]		No D[u]		No D[u]

*Data overlap with Ikin 1963 and with Bonné 1966.

	MOROCCO — Levene 1975, N = 421		MOROCCO — Bonné-Tamir 1976, N = 192		TUNISIA — Levene 1975, N = 98		AUSTRIA — N = 187		BULGARIA — Levene 1975, N = 283		GERMANY — N = 513		GERMANY — Bonné-Tamir 1976, N = 93		YUGOSLAVIA — Levene 1975, N = 105		RUSSIA — Levene 1975, N = 577	
	Obs.	Exp.	Obs.	Exp.	Obs.	Exp.	Obs.	Exp.	Obs.	Exp.	Obs.	Exp.	Obs.	Exp.	Obs.	Exp.	Obs.	Exp.
CCDEE		0·17		0·03				0·02		0	1	0·05		0	1	0·05		0
CCDEe	10	8·46	4	3·17			2	2·00	2	1·50	4	4·92	1	0·73	1	2·17		0·92
CCDee	106	101·71	65	58·67	30	27·57	49	47·24	66	73·78	129	122·24	23	21·99	29	22·05	155	148·35
CCddee		0				0·03		0·02		0·06		0		0·02				0
CcDEE	2	1·60		0·56			1	0·58		0·39	1	1·28		0·20		0·62	1	0·23
CcDEe	34	45·05	20	22·54	8	6·90	22	28·52	39	42·36	56	67·20	11	12·49	14	14·34	73	74·32
CcDee	161	161·79	60	71·15	35	40·92	66	63·36	115	96·70	184	186·42	32	32·85	25	37·46	202	212·85
CcddEe										0·06		0						
Ccddee	2	1·98	4	1·80	1	0·99	1	0·98	2	1·84	1	1·59	1	0·97			1	1·04
ccDEE	5	3·58				0·43	7	3·91	9	5·80	13	8·26	3	1·61	1	1·76	10	9·12
ccDEe	37	30·78	11	12·50	5	5·23	18	18·48	24	26·66	49	47·91	8	9·15	13	10·63	52	53·08
ccDee	23	23·70	15	11·56	7	5·85	10	10·41	10	13·22	28	27·49	2	1·86	10	7·58	32	29·72
ccddEE										0		0						
ccddEe									1	0·96		0·82						
ccddee	41	42·18	13	10·02	12	10·08	11	11·48	15	19·67	46	44·83	12	11·13	11	8·34	51	47·37
CDE	2·04		1·49				1·07		0·52		0·98		0·81		2·26		0·16	
CDe	48·41		55·28		51·48		49·21		49·84		48·29		47·14		45·83		50·40	
Cde	0·74				1·58		1·06		1·23		0·53		1·51				0·31	
cDE	9·24		9·71		6·63		14·44		13·67		12·39		13·17		12·98		12·58	
cDe	7·91		10·68		8·24		9·45		7·72		7·98		2·77		10·75		7·90	
cdE									0·65		0·27							
cde	31·66		22·84		32·07		24·77		26·37		29·56		34·60		28·18		28·65	
	No D^u		No D^u		No D^u		No D^u		No D^u		No D^u		No D^u		No D^u		No D^u	

Place: NORTHERN AFRICA (Morocco, Tunisia) — EUROPE (Austria, Bulgaria, Germany, Yugoslavia) — U.S.S.R. (Russia)

TABLE 4.8
RHESUS SYSTEM: TESTS WITH ANTI-C, -D, -E, -c, AND -e AND FOR D^u

SOUTH-WEST ASIA

	ISRAEL						IRAQ						YEMEN			
Place →	Born there		Nablus: Samaritans, Levi		Nablus: Samaritans, Ephraim											Saada, San'a, Damar, Beida, Aden	
Authors	Levene 1968		Ikin 1963		Ikin 1963		Levene 1975		Bonné-Tamir 1976		Dreyfuss et al. 1952		Levene 1975		Bodmer et al. 1972	
Number	1401		33(a)*		49(a)*		1146		185		104		373		202(a)	
	Obs.	Exp.	Obs.	Exp.	Obs.	Exp.	Obs.	Exp.	Obs.	Exp.	Obs.	Exp.	Obs.	Exp.	Obs.	Exp.
CCDEE	1	0·14					1	0·05		0				0·19		
CCDEe	15	14·62					4	8·27	1	0·81			7	8·10		
CCDee	363	373·80	7	7·30	18	14·73	338	331·71	66	63·75	26	25·41	88	82·84	63	55·03
CCD^{u}ee			2	3·07		0·14		0·06		0·28		0·09	1	0·05	1	0·08
CCddee	1	0·09					1	0·30		0·20			1	0·08		
CcDEE	2	3·55					5	2·57					4	1·79		
CcDEe	182	191·41	4	3·93	1	3·30	208	210·82	30	32·48	3	6·94	39	43·64	19	21·41
CcD^{u}Ee								0·01								
CcDee	527	497·08	6	5·62	15	18·89	339	344·57	52	53·62	45	42·41	132	136·75	64	76·86
CcD^{u}ee	1	0·09	9	6·71	2	2·05	1	0·92		0·19	3	2·64		1·23		2·53
CcddEe		0·09						0·12								
ccDEE	4	6·28		0·37	1	0·19	8	9·35	3	2·87		0·47	7	3·10	4	2·08
ccD^{u}EE	33	22·04							4	4·03				4·07		
ccDEe	108	119·16	3	2·33	3	2·33	115	108·57	17	14·22	11	6·13	27	31·22	14	15·43
ccD^{u}Ee		0·04														
ccDee	56	59·24					23	23·05	4	4·18	1	1·24	22	20·48	8	6·38
ccD^{u}ee	2	3·04						0·01	1	1·01	11	13·69				
ccddEE		0·02						1·78								
ccddEe	3	3·10					2									
ccddee	103	107·19	2	3·67	9	7·37	72	71·74	7	7·36	4	4·98	45	39·46	29	22·20
CDE	1·01						0·67		0·37				2·30			
CDe	50·85		25·54		49·68		52·04		54·94		46·52		45·38		50·41	
CD^{u}e	0·81		30·52		5·42		0·16				3·00		0·51		1·82	
Cde	12·15		10·61		6·12		1·63		3·89				1·28			
cDE							16·43		14·76		6·73		10·43		10·15	
cD^{u}E																
cDe	6·73						3·74		4·78		1·38		7·57		4·47	
cD^{u}c	0·39						0·31		1·32		20·48					
cdE	0·40															
cde	27·66		33·33		38·78		25·02		19·94		21·88		32·53		33·15	

*Data overlap with Bonné-Tamir 1975.

	AFRICA				EUROPE								TURKEY	
Place	EGYPT		ETHIOPIA†		CZECHOSLOVAKIA		HUNGARY		POLAND		ROMANIA		TURKEY	
Population			Falasha											
Authors	Levene 1975		Bat-Miriam 1962						Levene 1975					
Number	534		152		375		272		1306		1226		298	
	Obs.	Exp.	Obs.	Exp.	Obs.	Exp.	Obs.	Exp.	Obs.	Exp.	Obs.	Exp.	Obs.	Exp.
$CCDEE$	1	0·05				0	1	0		0		0		0·06
$CCDEe$	6	6·46			2	1·69		0·77	8	7·96	8	7·97	5	4·17
$CCDee$	140	147·33	12	9·21	101	104·36	76	75·45	338	341·91	339	336·29	81	81·18
CCD^uee						0·04		0·01		0		0		
$CCddee$	1	0·37		0·03		0·04		0·19		0				
$CcDEE$		1·50				0·41		0·19	2	1·96	2	1·96		0·89
$CcDEe$	78	73·42	3	3·22	56	53·29	30	38·13	181	181·80	174	172·01	37	36·24
CcD^uEe										0				
$CcDee$	193	183·00	47	52·38	134	130·13	103	96·12	472	463·37	426	433·76	110	111·15
CcD^uee		0·11		0·10	1	0·97		0·03	1	1·04				
$CcddEe$										0		0		
$ccdEE$	7	7·48	1	0·84	2	1·95	1	1·03	4	4·05	3	3·06	4	3·37
ccD^uEE	8	8·12	3	0·27	6	6·52	7	4·71	25	22·72	19	20·72		0·04
$ccDEe$	42	44·70	4	9·20	32	33·22	28	24·24	114	119·11	108	106·54	20	20·86
ccD^uEe													1	0·98
$ccDee$	20	21·25	72	67·38	17	17·55	7	8·43	47	47·41	56	54·80	19	18·55
ccD^uee	1	1·01	2	1·88			1	1·21		0	1	0·86		
$ccddEE$									1			0		0·01
$ccddEe$									1	1·57	2	1·96	1	0·98
$ccddee$	37	39·20	8	7·48	24	24·83	18	21·68	112	113·10	88	86·07	20	19·52
CDE		1·15				0·42		0·27		0·59		0·62		1·34
CDe		50·00		23·39		51·25		51·99		50·50		51·90		52·19
CD^ue						0·51		0·68		0·14		0·47		
Cde		2·59		1·25		1·01				0·53				9·43
cDE		12·33		4·26		13·19		13·14		12·99		12·69		0·65
cD^uE														
cDe		6·48		46·24		7·90		4·92		5·63		7·37		10·14
cD^ue		0·36		2·63				0·77		0·20		0·14		0·65
cdE												0·31		
cde		27·09		22·18		25·72		28·23		29·42		26·50		25·60

TABLE 4.9

RHESUS SYSTEM: TESTS WITH ANTI-C, -D, -E, -c, -e, AND -V

Place	IRAQ				IRAN		YEMEN			
	N.W.		S.E.				N. and N.E.		S.	
Authors	Tills *et al.* 1977									
Population	*Kurdish*		*Kurdish*		*Kurdish*					
Number	61		50		106		65		92	
	Obs.	Exp.	Obs.	Exp.	Obs.	Exp.	Obs.	Exp.	Obs.	Exp.
CCDEE		0·15		0·28		0·23		0·03		0·01
CCDEe		2·87		4·83	5	5·44		1·56		0·69
CCDee	13	14·26	24	20·81	31	31·12	24	24·26	25	22·77
CCddee		0·03				0·06				0·04
CcDEE	4	1·12	7	1·54	3	2·17	2	0·19	1	0·22
CcDEeV								0·04		0·10
CcDEe	15	12·82	12	13·79	28	26·88	5	6·48	15	15·11
CcDeeV								1·24	1	6·38
CcDee	19	15·67	5	4·83	21	21·28	27	22·33	25	21·25
Ccddee	1	0·76			1	0·98			1	0·93
ccDEE	1	2·18	1	2·11	6	4·98		0·35	2	2·38
ccDEeV							1	0·14	4	2·06
ccDEe	4	6·42		1·53	5	8·89	2	2·65	6	7·82
ccDeeV							1	0·59	7	3·83
ccDee									1	1·29
ccddee	4	4·72	1	0·28	6	3·97	3	5·14	4	5·12

	N.W.	S.E.	IRAN	N. and N.E.	S.
CDE	4·86	7·49	4·73	1·98	0·75
CDe	46·17	64·51	51·86	61·10	47·64
Cde	2·24		2·37		2·15
cDE	18·91	20·51	21·69	7·25	16·10
cDeV				1·56	6·97
cDe					2·81
cde	27·82	7·49	19·35	28·11	23·58

TABLE 4.10
RHESUS SYSTEM: TESTS WITH ANTI-C, -D, -E, -c, -e, AND -V AND FOR D^u

	SOUTH-WEST ASIA		AFRICA	
Place	YEMEN **Habban**		MOROCCO **Tafilalet Oases†**	
Authors	Bonné *et al.* 1970		Ikin *et al.* 1972	
Number	596		146	
	Obs.	Exp.	Obs.	Exp.
CCDEE			0·22	0·20
CCDEe			5·78	5·50
CCDee	126	122·12	36	34·24
CCD^{u}ee	3	0·60	1	3·26
CCddee		0·60		0·09
CcDEE			0·62	0·67
CcDEcV				0·61
CcDEe	6	6·32	7·38	12·80
CcDeeV	106	124·86	10	7·72
CcDee	162	151·62	42	36·41
CcD^{u}eeV			0·11	0·60
CcD^{u}ee	1	3·64	10·89	11·42
Ccddee	9	9·00	2	2·19
ccDEE		0·06	2	0·55
ccDEeV	4	3·22	1	1·01
ccDEe	4	4·23	9	6·09
ccDeeV	129	114·85	4	4·75
ccDee	17	21·04	2	2·80
ccD^{u}eeV			1	1·26
ccD^{u}ee	1	1·13		
ccddee	28	32·71	11	13·83
CDE			3·71	
CDe	41·01		35·60	
CD^{u}e	1·24		12·71	
Cde	3·23		2·43	
cDE	1·17		6·17	
cDeV	23·03		4·25	
cDe	6·53		2·98	
cD^{u}eV			1·37	
cD^{u}e	0·37			
cde	23·42		30·78	

All C + D + not tested for e;
1 CcD^{u}ee not tested for V.

TABLE 5
LUTHERAN SYSTEM: TESTS WITH ANTI-Lu^a

Place	Population	Authors	Number	Lu(a+)	Lu(a−)	Lu^a	Lu^b
ISRAEL, Nablus	Samaritans, Levi	Ikin 1963	33(a)	2	31	3·08	96·92
ISRAEL, Nablus	Samaritans, Ephraim	Ikin 1963	49(a)	16	33	17·93	82·07
ISRAEL, Nablus, Jaffa	Samaritans	Bonné-Tamir 1975*	109	1	108	0·46	99·54
IRAQ, N.W.	Kurdish	Tills et al. 1977	61	1	60	0·82	99·18
IRAQ, S.E.	Kurdish	Tills et al. 1977	50	1	49	1·01	98·99
IRAN	Kurdish	Tills et al. 1977	106	1	105	0·47	99·53
YEMEN, N. & N.E.		Tills et al. 1977	65	1	64	0·77	99·23
YEMEN, S.		Tills et al. 1977	92	1	91	0·54	99·46
YEMEN, Habban		Bonné et al. 1970	559		559		100·00
LIBYA		Bonné-Tamir 1974	130	14	116	5·54	94·46
GERMANY		Bonné-Tamir 1975	67	5	62	3·80	96·20

*Data overlap with Ikin 1963.

TABLE 6.1
KELL SYSTEM: TESTS WITH ANTI-K

Place	Population	Authors	Number	K+	K−	*K*	*k*
SOUTH-WEST ASIA							
ISRAEL		Levene 1975	1399	164	1235	6·04	93·96
ISRAEL, Jaffa, Nablus	Samaritans	Bonné-Tamir 1975	124*		124		100·00
ISRAEL, Jaffa	Samaritans	Bonné 1966	132		132		100·00
LEBANON, SYRIA		Levene 1975	203	14	189	3·51	96·49
IRAQ		Levene 1975	1146	136	1008	6·21	93·79
IRAQ, Hit	Karaites	Cohen 1971	72	13	59	9·48	90·52
IRAN		Levene 1975	152	35	117	12·27	87·73
YEMEN		Levene 1975	374	15	359	2·03	97·97
YEMEN, Habban		Bonné *et al.* 1970	580	5	575	0·43	99·57
YEMEN, Saada, San'a, Damar, Beida, Aden		Bodmer *et al.* 1972	202(a)	7	195	1·75	98·25
ADEN		Levene 1975	127	8	119	3·20	96·80
AFRICA							
EGYPT, SUDAN		Levene 1975	533	45	488	4·31	95·69
LIBYA		Levene 1975	99	11	88	5·72	94·28
LIBYA		Bonné-Tamir 1974	148	9	139	3·09	96·91
MOROCCO		Levene 1975	423	52	371	6·35	93·65
MOROCCO, Tafilalet Oases†		Ikin *et al.* 1972	145	17	128	6·04	93·96
TUNISIA		Levene 1975	98	5	93	2·58	97·42
TUNISIA, Île de Djerba†		Moullec & Abdelmoula 1954	27		27		100·00
EUROPE							
AUSTRIA		Levene 1975	187	22	165	6·06	93·94
BULGARIA		Levene 1975	283	17	266	3·05	96·95
CZECHOSLOVAKIA		Levene 1975	378	59	319	8·14	91·86
GERMANY		Levene 1975	513	74	439	7·49	92·51
GERMANY		Bonné-Tamir 1975	93	13	80	7·25	92·75
HUNGARY		Levene 1975	271	43	228	8·28	91·72
POLAND		Levene 1975	1304	163	1141	6·46	93·54
ROMANIA		Levene 1975	1227	164	1063	6·92	93·08
YUGOSLAVIA		Levene 1975	105	12	93	5·89	94·11
U.S.S.R.							
RUSSIA		Levene 1975	577	60	517	5·34	94·66
TURKEY		Levene 1975	297	28	269	4·83	95·17
AMERICA							
CANADA		Chown & Lewis 1951	80	9	71	5·79	94·21

*Data overlap with Ikin *et al.* 1963 and with Bonné 1966.

TABLE 6.2
KELL SYSTEM: TESTS WITH ANTI-K AND -k

Place	Population	Authors	Number	KK	Kk	kk	*K*	*k*
IRAQ		Bonné-Tamir 1976	184	1	18	165	5·43	94·57
IRAQ, N.W.	Kurdish	Tills *et al.* 1977	61		4	57	3·28	96·72
IRAQ, S.E.	Kurdish	Tills *et al.* 1977	50		1	49	1·00	99·00
IRAN	Kurdish	Tills *et al.* 1977	106		5	101	2·36	97·64
IRAN		Levene *et al.* 1977	159	1	36	120	13·21	86·79
YEMEN, N. & N.E.		Tills *et al.* 1977	65		2	63	1·54	98·46
YEMEN, S.		Tills *et al.* 1977	92		1	91	0·54	99·46
MOROCCO		Bonné-Tamir 1976	191		22	169	5·76	94·24

TABLE 6.3
KELL SYSTEM: TESTS WITH ANTI-Ul^a (KARHULA)

Place	Population	Authors	Number Ul(a−)
FINLAND† Israeli, non-Ashkenazi	Furuhjelm *et al.* 1968	13

TABLE 6.4
KELL SYSTEM: TESTS WITH ANTI-Kp^a AND -Kp^b; ALL Kp(a−b+)

Place	Population	Authors	Number	Kp(a−b+)
ISRAEL, Jaffa	Samaritans	Bonné 1966	132	132

TABLE 6.5
KELL SYSTEM: TESTS WITH ANTI-Js^a (SUTTER)

Place	Population	Authors	Number	Js(a+)	Js(a−)	*Js^a*	*Js^b*
IRAQ, N.W.	Kurdish	Tills *et al.* 1977	20		20		100·00
IRAQ, S.E.	Kurdish	Tills *et al.* 1977	9		9		
IRAN	Kurdish	Tills *et al.* 1977	87		87		100·00
YEMEN, N. & N.E.		Tills *et al.* 1977	65		65		100·00
YEMEN, S.		Tills *et al.* 1977	92	2	90	1·09	98·91
YEMEN, Habban		Bonné *et al.* 1970	578	75	503	6·72	93·28

TABLE 6.6
KELL SYSTEM: TESTS WITH ANTI-K AND -Js^a; ALL K−Js(a−)

Place	Population	Authors	Number	K−Js(a−)
ISRAEL, Nablus	Samaritans	Ikin *et al.* 1963*	81(a)	81

*Data overlap with Bonné-Tamir 1975.

TABLE 7.1

ABH SECRETION AND LEWIS SYSTEMS: SALIVA TESTED FOR ABH SECRETION

Place	Population	Authors	Number	Secretors	Non-Secretors	*Se*	*se*
MIDDLE EAST		Micle *et al.* 1977	97	68	29	45·32	54·68
ISRAEL, Jaffa	Samaritans	Bonné 1966	118	71	47	36·89	63·11
YEMEN, Habban		Bonné *et al.* 1970	538(c)	413	125	51·80	48·20
NORTH AFRICA		Micle *et al.* 1977	132	97	35	48·51	51·49
EASTERN and CENTRAL EUROPE		Micle *et al.* 1977	348	269	79	52·35	47·65
U.S.A., New York City†		Schiff 1940	74	61	13	58·08	41·92

TABLE 7.2

ABH SECRETION AND LEWIS SYSTEMS: RED CELLS TESTED WITH ANTI-Le[a]

Place	Population	Authors	Number	Le(a+)	Le(a−)	Le(a+)%
TUNISIA, Île de Djerba†		Moullec & Abdelmoula 1954	33	8	25	24·24

TABLE 8.1
DUFFY SYSTEM: TESTS WITH ANTI-Fy[a]

Place	Population	Authors	Number	Fy(a+)	Fy(a−)	Fy^a	$Fy^b + Fy$
SOUTH-WEST ASIA							
ISRAEL		Levene 1975	1399	892	507	39·80	60·20
ISRAEL, Jaffa	Samaritans	Bonné 1966*	128	92	36	46·97	53·03
ISRAEL, Nablus	Samaritans, Levi	Ikin et al. 1963*	33(a)	22	11	42·27	57·73
ISRAEL, Nablus	Samaritans, Ephraim	Ikin et al. 1963*	48(a)	19	29	22·27	77·73
LEBANON, SYRIA		Levene 1975	203	134	69	41·70	58·30
IRAQ		Levene 1975	1146	772	374	42·87	57·13
IRAQ, Hit	Karaites	Goldschmidt et al. 1976	72	55	17	51·41	48·59
IRAQ, N.W.	Kurdish	Tills et al. 1977	61	43	18	45·68	54·32
IRAQ, S.E.	Kurdish	Tills et al. 1977	50	32	18	40·00	60·00
IRAN	Kurdish	Tills et al. 1977	106	83	23	53·42	46·58
IRAN		Levene 1975	152	110	42	47·44	52·56
IRAN		Levene et al. 1977	158	100	58	39·41	60·59
YEMEN		Levene 1975	374	156	218	23·65	76·35
YEMEN, N. and N.E.		Tills et al. 1977	65	16	49	13·18	86·82
YEMEN, S.		Tills et al. 1977	92	36	56	21·98	78·02
ADEN		Levene 1975	127	70	57	33·01	66·99
NORTHERN AFRICA							
EGYPT, SUDAN		Levene 1975	533	313	220	35·75	64·25
LIBYA		Levene 1975	99	64	35	40·54	59·46
MOROCCO		Levene 1975	423	248	175	35·68	64·32
MOROCCO, Tafilalet Oases†		Ikin et al. 1972	145	72	73	29·05	70·95
TUNISIA		Levene 1975	98	58	40	36·11	63·89
EUROPE							
AUSTRIA		Levene 1975	187	131	56	45·27	54·73
BULGARIA		Levene 1975	283	189	94	42·36	57·64
CZECHOSLOVAKIA		Levene 1975	378	251	127	42·03	57·97
GERMANY		Levene 1975	512	339	173	41·87	58·13
HUNGARY		Levene 1975	271	185	86	43·67	56·33
POLAND		Levene 1975	1305	903	402	44·50	55·50
ROMANIA		Levene 1975	1226	826	400	42·88	57·12
YUGOSLAVIA		Levene 1975	105	82	23	53·20	46·80
U.S.S.R.							
RUSSIA		Levene 1975	577	401	176	44·77	55·23
TURKEY		Levene 1975	297	196	101	41·68	58·32

*Data overlap with Bonné-Tamir 1975.

TABLE 8.2
DUFFY SYSTEM: TESTS WITH ANTI-Fy^a AND -Fy^b

Place	Population	Authors	Number	Fy(a+b−)	Fy(a+b+)	Fy(a−b+)	Fy(a−b−)	Fy^a	Fy^b	Fy	χ²
ISRAEL, Jaffa, Nablus	Samaritans	Bonné-Tamir 1975	124*	31	53	40		46·37	53·63		2·46
IRAQ		Bonné-Tamir 1976	184	44	89	51		48·10	51·90		0·18
IRAQ, N.W.	Kurdish	Tills et al. 1977	20**	9	6	4	1	49·59	29·05	21·36	0·03
IRAQ, S.E.	Kurdish	Tills et al. 1977	8**	1	3	3	1				
IRAN	Kurdish	Tills et al. 1977	87**	31	38	14	4	51·10	34·28	14·62	6·23
YEMEN, N. & N.E.		Tills et al. 1977	22**	6	4	2	10	25·07	14·11	60·82	5·94
YEMEN, S.		Tills et al. 1977	51**	11	1	4	35	12·53	5·02	82·45	0·24
YEMEN, Habban		Bonné et al. 1970	496	175	43	109	169	25·11	16·70	58·19	0·07
YEMEN, Saada, San'a, Damar, Beida, Aden		Bonné-Tamir 1977	168	39	27	62	40	21·82	31·07	47·11	1·45
LIBYA		Bonné-Tamir 1976	148	36	61	51		44·93	55·07		4·13
MOROCCO		Bonné-Tamir 1976	191	24	63	104		29·06	70·94		7·65
GERMANY		Bonné-Tamir 1976	93	20	37	36		41·40	58·60		3·01

* Data overlap with Ikin et al. 1963 and with Bonné 1966.
**Are included in the respective samples tested with anti-Fy^a only.

TABLE 9
KIDD SYSTEM: TESTS WITH ANTI-Jk^a

Place	Population	Authors	Number	Jk(a+)	Jk(a−)	Jk^a	$Jk^b + Jk$
SOUTH-WEST ASIA							
ISRAEL		Levene 1975	1398	1117	281	55·17	44·83
ISRAEL, Nablus, Jaffa	Samaritans	Bonné-Tamir 1975	124*	92	32	49·20	50·80
ISRAEL, Jaffa	Samaritans	Bonné 1966	132	92	40	44·95	55·05
LEBANON, SYRIA		Levene 1975	203	162	41	55·06	44·94
IRAQ		Levene 1975	1145	932	213	56·87	43·13
IRAQ		Bonné-Tamir 1976	182	114	68	38·88	61·12
IRAQ, N.W.	Kurdish	Tills et al. 1977	20	11	9	32·92	67·08
IRAQ, S.E.	Kurdish	Tills et al. 1977	9	7	2		
IRAN	Kurdish	Tills et al. 1977	87	60	27	44·30	55·70
IRAN		Levene 1975	152	131	21	62·82	37·18
YEMEN		Levene 1975	373	299	74	55·46	44·54
YEMEN, N. & N.E.		Tills et al. 1977	22	18	4	57·36	42·64
YEMEN, S.		Tills et al. 1977	51	46	5	68·69	31·31
YEMEN, Habban		Bonné et al. 1970	577	499	78	63·23	36·77
YEMEN, Saada, San'a, Damar, Beida, Aden		Bonné 1977	168	126	42	50·00	50·00
ADEN		Levene 1975	127	109	18	62·36	37·64
AFRICA							
EGYPT, SUDAN		Levene 1975	533	431	102	56·25	43·75
LIBYA		Levene 1975	99	77	22	52·86	47·14
LIBYA		Bonné-Tamir 1974	148	90	58	37·40	62·60
MOROCCO		Bonné-Tamir 1976	183	118	65	40·40	59·60
MOROCCO		Levene 1975	423	337	86	54·91	45·09
TUNISIA		Levene 1975	98	70	28	46·55	53·45
EUROPE							
AUSTRIA		Levene 1975	187	165	22	65·71	34·29
BULGARIA		Levene 1975	283	236	47	59·24	40·76
CZECHOSLOVAKIA		Levene 1975	376	316	60	60·05	39·95
GERMANY		Levene 1975	511	440	71	62·73	37·27
GERMANY		Bonné-Tamir 1974	71	50	21	45·61	54·39
HUNGARY		Levene 1975	271	231	40	61·58	38·42
POLAND		Levene 1975	1305	1092	213	59·60	40·40
ROMANIA		Levene 1975	1226	989	237	56·03	43·97
YUGOSLAVIA		Levene 1975	105	87	18	58·60	41·40
U.S.S.R.							
RUSSIA		Levene 1975	577	488	89	60·73	39·27
TURKEY		Levene 1975	297	229	68	52·15	47·85

*Data overlap with Bonné 1966.

TABLE 10
DIEGO SYSTEM: TESTS WITH ANTI-Di[a]

Place	Population	Authors	Number	Di(a+)	Di(a−)	Di^a	Di^b
SOUTH-WEST ASIA							
ISRAEL, Jaffa	Samaritans	Bonné 1966	132		132		100·00
ISRAEL, Nablus	Samaritans	Ikin *et al.* 1963	81(a)		81		100·00
YEMEN		Godber *et al.* 1973	75		75		100·00
YEMEN, Habban		Bonné *et al.* 1970	101		101		100·00
IRAN and IRAQ	Kurdish	Godber *et al.* 1973	116		116		100·00
INDIA							
KERALA, Cochin	'Black'	Gurevitch 1965	45		45		100·00
AMERICA							
ARGENTINE†	Ashkenazi	Palatnik 1965	200		200		100·00
ARGENTINE†	Sephardi	Palatnik 1965	200	6	194	1·51	98·49
BRAZIL, São Paulo†	Ashkenazi	Ottensooser *et al.* 1962	100		100		100·00

TABLE 11
VARIOUS BLOOD GROUP SYSTEMS

System	Place	Population	Authors	Number				
Dombrock				Number	Do(a+)	Do(a−)	Do^a	Do
	ISRAEL		Tippett 1967	128	83	45	40·70	59·30
Radin				Number	Rd+	Rd−	Rd+%	
	ISRAEL, Jaffa	Samaritans	Bonné 1966	132		132		
	ISRAEL, Jerusalem		Lundsgaard & Jensen 1971	4501	18	4483	0·40	
	U.S.A., New York†		Rausen *et al.* 1967	562	3	559	0·53	
Wright				Number	Wr(a+)			
	ISRAEL, Jaffa	Samaritans	Bonné 1966	132	0			
	YEMEN		Godber *et al.* 1973	75	0			
	YEMEN, Habban		Bonné *et al.* 1970	596	0			
	IRAQ and IRAN	Kurdish	Godber *et al.* 1973	116	0			
Gonzales				Number	Go(a+)			
	U.S.A., New York†	Donors, patients	Alter *et al.* 1967	200	0			
Stoltzfus				Number	Sf(a+)	Sf(a−)	Sf^a	Sf
	SYRIA	Sampled in New York	Bias *et al.* 1969	307	101	206	18·09	81·91

TABLE 12
Xg SYSTEM

Place	Population	Authors	Total	Males Number	Males Xg(a+)	Males Xg(a−)	Females Number	Females Xg(a+)	Females Xg(a−)	Xg^a	Xg	χ² Males	χ² Females
	North African and Oriental	Adam et al. 1967	201	106	69	37	95	87	8	67·77	32·23	0·35	0·35

TABLE 13
HAPTOGLOBIN SYSTEM

Place	Population	Authors	Number	Hp1-1	Hp2-1	Hp2-2	Hp2-1M	Hp2-1M%	Hp0	Hp0%	Hp^1	Hp^2	χ²
ISRAEL	Unclassified	Fried et al. 1963	175	20	69	86					31·14	68·86	1·14
ISRAEL	Ashkenazi	Ramot et al. 1961	170	13	88	69					33·53	66·47	4·42
ISRAEL	Ashkenazi	Fried et al. 1963	669	56	290	320			3	0·45	30·18	69·82	0·74
ISRAEL	Sephardi	Fried et al. 1963	44	5	23	16					37·50	62·50	0·58
ISRAEL, Jaffa	Samaritans	Bonné 1966	125	20	59	46					39·60	60·40	0·02
ISRAEL, Nablus	Samaritans	Bonné et al. 1967	37	4	17	2			4	10·81	37·88	62·12	0·30
NEAR EAST		Fried et al. 1963	48	2	21	25					26·04	73·96	0·88
IRAQ		Ramot et al. 1961	118	9	50	59					28·81	71·19	0·13
IRAQ		Levene et al. 1977a	142	10	49	83					24·30	75·70	0·55
IRAQ, Baghdad		Fried et al. 1963	197	14	79	103	1	0·51			27·41	72·59	0·08
IRAQ, Hit	Karaites	Goldschmidt et al. 1976	69	3	33	33					28·26	71·74	2·23
IRAQ, N.W.	Kurdish	Tills et al. 1977	61	7	28	26					34·43	65·57	0·02
IRAQ, S.E.	Kurdish	Tills et al. 1977	50	4	19	27					27·00	73·00	0·07
IRAN	Kurdish	Tills et al. 1977	106	9	35	61			1	0·94	25·24	74·76	1·43
IRAN		Ramot et al. 1962	91	9	37	45					30·22	69·78	0·12
IRAN		Fried et al. 1962	101	8	43	50					29·21	70·79	0·11
KURDISTAN		Ramot et al. 1962	113	12	57	42			2	1·77	36·49	63·51	1·29
KURDISTAN		Fried et al. 1963	96	5	48	41			2	2·08	30·85	69·15	3·65
YEMEN		Fried et al. 1963	41	2	16	22			1	2·44	25·00	75·00	0·18
YEMEN, N. & N.E.		Tills et al. 1977	65	5	38	22					36·92	63·08	4·23
YEMEN, S.		Tills et al. 1977	90	5	36	49					25·56	74·44	0·20
YEMEN, Habban		Bonné et al. 1970	589	26	197	358			8	1·36	21·43	78·57	0·03
YEMEN, Saada, San'a, Damar, Beida, Aden		Bodmer et al. 1972	202(a)	19	96	87					33·17	66·83	1·04
NORTHERN AFRICA		Ramot et al. 1961	104	8	43	52	1	0·96			28·85	71·15	0·10
NORTHERN AFRICA		Fried et al. 1963	223	17	90	113	1	0·45	2	0·90	28·28	71·72	0·05
LIBYA		Bonné-Tamir 1974	148	8	55	85					23·99	76·01	0·05
POLAND		Levene et al. 1977a	105	15	49	41					37·62	62·38	<0·01

TABLE 14
TRANSFERRIN SYSTEM

Place	Population	Authors	Number	TfC	TfBC	TfB	Tf^C	Tf^B
ISRAEL	Mixed	Ramot *et al.* 1962	596	596			100·00	
ISRAEL	Mixed	Fried *et al.* 1963	1594	1594			100·00	
ISRAEL, Nablus	Samaritans	Bonné *et al.* 1967	37	37			100·00	
IRAQ, N.W.	Kurdish	Tills *et al.* 1977	61	59	2		98·36	1·64
IRAQ, S.E.	Kurdish	Tills *et al.* 1977	50	50			100·00	
IRAN	Kurdish	Tills *et al.* 1977	106	106			100·00	
YEMEN, N. & N.E.		Tills *et al.* 1977	65	65			100·00	
YEMEN, S.		Tills *et al.* 1977	92	92			100·00	
YEMEN, Habban		Bonné *et al.* 1970	598	595	3		99·75	0·25
YEMEN, Saada, San'a, Damar, Beida, Aden		Bodmer *et al.* 1972	202	202			100·00	

TABLE 15
Gc SYSTEM

Place	Population	Authors	Number	Gc1-1	Gc2-1	Gc2-2	Gc^1	Gc^2	χ^2
SOUTH-WEST ASIA									
ISRAEL	Ashkenazi	Cleve *et al.* 1962	99	45	41	13	66·16	33·84	0·56
ISRAEL, Jaffa	Samaritans	Bonné 1966	125	79	41	5	79·60	20·40	0·01
ISRAEL, Nablus	Samaritans	Bonné *et al.* 1967	9	6	3				
IRAQ		Cleve *et al.* 1962	85	48	33	4	75·88	24·12	0·31
IRAQ		Levene *et al.* 1977a	141	106	27	8	84·75	15·25	9·51
IRAN		Cleve *et al.* 1962	49	30	14	5	75·51	24·49	2·53
IRAN		Kitchin & Bearn 1964	149	91	43	15	75·50	24·50	7·21
KURDISTAN		Levene *et al.* 1977a	68	45	16	7	77·94	22·06	6·81
KURDISTAN, IRAQ and IRAN		Cleve *et al.* 1962	42	27	14	1	80·95	19·05	0·27
YEMEN		Cleve *et al.* 1962	49	30	19	0	80·61	19·39	2·83
INDIA									
		Kitchin & Bearn 1964	71	37	28	6	71·83	28·17	0·05
KERALA, Cochin		Kitchin & Bearn 1964	53	32	19	2	78·30	21·70	0·16
NORTHERN AFRICA									
ALGERIA, LIBYA, MOROCCO, TUNISIA		Cleve *et al.* 1962	64	31	28	5	70·31	29·69	0·15
ALGERIA, LIBYA, MOROCCO, TUNISIA		Kitchin & Bearn 1964	109	54	49	6	72·02	27·98	1·45
EUROPE									
POLAND		Levene *et al.* 1977a	103	59	33	11	73·30	26·70	3·40

TABLE 16.1
Ag (BETA LIPOPROTEIN) SYSTEM: TESTS WITH C. de B. SERUM

Place	Population	Authors	Number	Ag(+)	Ag(−)	Ag(+)%
IRAN		Blumberg *et al.* 1963	37	26	11	70·27
KURDISTAN		Blumberg *et al.* 1963	17	12	5	70·59

TABLE 16.2
Ag (BETA LIPOPROTEIN) SYSTEM: TESTS WITH ANTI-Ag^x

Place	Population	Authors	Number	Ag(x+)	Ag(x−)	Ag^x	Ag^y
IRAQ, N.W.	Kurdish	Tills *et al.* 1977	41	21	20	30·16	69·84
IRAN	Kurdish	Tills *et al.* 1977	19	12	7		
YEMEN, N. & N.E.		Tills *et al.* 1977	43	10	33	12·40	87·60
YEMEN, S.		Tills *et al.* 1977	41	8	33	10·29	89·71

TABLE 17
Lp (BETA LIPOPROTEIN) SYSTEM: TESTS WITH ANTI-Lp^a

Place	Population	Authors	Number	Lp(a+)	Lp(a−)	Lp^a	Lp
IRAQ, N.W.	Kurdish	Tills *et al.* 1977	41	14	27	18·85	81·15
IRAQ, S.E.	Kurdish	Tills *et al.* 1977	41	20	21	28·43	71·57
IRAN	Kurdish	Tills *et al.* 1977	19	12	7		
YEMEN, N. & N.E.		Tills *et al.* 1977	43	9	34	11·08	88·92
YEMEN, S.		Tills *et al.* 1977	41	21	20	30·16	69·84

TABLE 18.1
PSEUDOCHOLINESTERASE E$_1$ SYSTEM: COMMON TYPES ONLY PRESENT

Place	Population	Authors	Number	U	I	A	E_1^u	E_1^a
ISRAEL		Szeinberg *et al.* 1966	244	239	5		98·98	1·02
ISRAEL	Males over 40	Szeinberg *et al.* 1972	1652	1583	69		97·91	2·09
ISRAEL	Ashkenazi	Szeinberg *et al.* 1966	923	893	29	1	98·32	1·68
LEBANON and SYRIA	Males over 40	Szeinberg *et al.* 1972	203	196	7		98·28	1·72
IRAN		Szeinberg *et al.* 1972	159	138	18	3	92·45	7·55
IRAN and IRAQ		Szeinberg *et al.* 1966	381	343	37	1	94·88	5·12
YEMEN		Szeinberg *et al.* 1966	124	116	7	1	96·37	3·63
MOROCCO		Kattamis *et al.* 1962	51	50	1		99·02	0·98
BALKANS and TURKEY		Szeinberg *et al.* 1966	214	207	7		98·36	1·64
BALKANS and TURKEY	Males over 40	Szeinberg *et al.* 1972	674	637	36	1	97·18	2·82

TABLE 18.2
PSEUDOCHOLINESTERASE E$_1$ SYSTEM: RARE TYPES PRESENT

Place	Population	Authors	Number	U	I	A	UF	IF	S	E_1^u	E_1^a	E_1^f	E_1^s
ISRAEL	Ashkenazi males over 40	Szeinberg *et al.* 1972	4196	4050	143			3		98·22	1·74	0·04	
IRAQ	Males over 40	Szeinberg *et al.* 1972	1057	960	92	3		2		95·18	4·73	0·09	
YEMEN	Males over 40	Szeinberg *et al.* 1972	459	441	17			1		97·93	1·96	0·11	
NORTHERN AFRICA	Males over 40	Szeinberg *et al.* 1972	1106	1072	30	1		2	1	95·46	1·45	0·09	3·00

TABLE 19
PSEUDOCHOLINESTERASE E$_2$ SYSTEM

Place	Population	Authors	Number	C$_5$+
ISRAEL, Nablus	Samaritans	Bonné et al. 1967	35	0
IRAQ		Robson & Harris 1966	64	0

TABLE 20
ACID PHOSPHATASE SYSTEM

Place	Population	Authors	Number	A	BA	B	CA	CB	C	P^a	P^b	P^c
ISRAEL	Ashkenazi	Goldschmidt 1967	479	35	161	255	8	20		24·95	72·13	2·92
ISRAEL	Samaritans	Bonné et al. 1967	37		5	32				6·76	93·24	
IRAQ		Goldschmidt 1967	82	11	33	35	1	2		34·15	64·02	1·83
IRAQ, Hit	Karaites	Goldschmidt et al. 1976	72			67		5			96·53	3·47
IRAQ, N.W.	Kurdish	Tills et al. 1977	61	10	30	16		5		40·98	54·92	4·10
IRAQ, S.E.	Kurdish	Tills et al. 1977	50	6	21	23				33·00	67·00	
IRAN	Kurdish	Tills et al. 1977	106	18	41	39	2	6		37·27	58·96	3·77
IRAQ and IRAN	Kurdish	Goldschmidt 1967	131	19	53	55	1	3		35·11	63·36	1·53
IRAN		Goldschmidt 1967	49	10	16	21		2		36·74	61·22	2·04
YEMEN		Goldschmidt 1967	37	1	9	26		1		14·87	83·78	1·35
YEMEN, N. & N.E.		Tills et al. 1977	23		2	17		4		4·35	86·96	8·69
YEMEN, S.		Tills et al. 1977	88	3	25	47	5	8		20·45	72·16	7·39
YEMEN, Habban		Bonné et al. 1970	314		13	284		17		2·07	95·22	2·71
YEMEN, Saada, San'a, Damar, Beida, Aden		Bodmer et al. 1972	199(a)	11	31	136	1	20		13·57	81·15	5·28
NORTHERN AFRICA		Goldschmidt 1967	137	5	47	76	2	6		21·69	75·37	2·94
LIBYA		Bonné-Tamir 1974	148	12	52	78	2	4		26·35	71·62	2·03
GERMANY		Bonné-Tamir 1975	92	7	38	44	2	1		29·35	69·02	1·63

TABLE 21.1.1

GLUCOSE-6-PHOSPHATE DEHYDROGENASE SYSTEM: NORMAL AND DEFICIENT TYPES IN MALES

Place	Population	Authors	Number	Normal	Deficient	Gd^-	
	Ashkenazi	Sheba *et al.* 1961	819	816	3	0·37	
	Karaites	Sheba *et al.* 1961	18	18			
	Samaritans	Szeinberg 1963	69	69			
BULGARIA, GREECE	Sephardi	Sheba *et al.* 1961	152(c)	151	1	0·66	
TURKEY	Sephardi	Sheba *et al.* 1961	256(c)	251	5	1·95	
Other than BULGARIA, GREECE, TURKEY	Sephardi	Sheba *et al.* 1961	93(c)	91	2	2·15	
SOUTH-WEST ASIA							
LEBANON and SYRIA		Szeinberg 1963	80(c)	75	5	6·25	
IRAQ		Sheba *et al.* 1961	902(c)	678	224	24·83	
IRAQ, Arbil, Kirkuk, Mosul		Szeinberg 1963	34		52·00		} Included in 902
IRAQ, Baghdad		Szeinberg 1963	286(c)	216	70	24·48	} by Sheba *et al.* 1961
IRAQ and IRAN	Kurdish	Sheba *et al.* 1961	196(c)	82	114	58·16	
IRAQ and IRAN, Arbil, Kirkuk, Mosul, Qasar-I-Shrin, Sulaymaniyah, Kermanshah, Sanandaj	Kurdish	Szeinberg 1963	59(c)	38	21	35·59	} Included in 196 by Sheba *et al.* 1961
IRAQ and IRAN, Amadiyah, Dahok, Sandor, Zakho	Kurdish	Szeinberg 1963	126(c)	37	89	70·63	
IRAN		Sheba *et al.* 1961	557(c)	473	84	15·08	
IRAN, Esfahan, Shiraz, Teheran		Szeinberg 1963	370(c)	330	40	10·81	} Included in 557
IRAN, Kermanshah, Sanandaj		Szeinberg 1963	45(c)	25	20	44·44	} by Sheba *et al.* 1961
IRAN, Esfahan†		Beaconsfield *et al.* 1967				14	
IRAN, Kermanshah†		Beaconsfield *et al.* 1967				29	
IRAN, Teheran†		Hedayat *et al.* 1969	108(c)	95	13	12·04	
AFGHANISTAN		Sheba *et al.* 1961	29(c)	26	3	10·34	
YEMEN		Sheba *et al.* 1961	415(c)	393	22	5·30	
INDIA	Bene Israel	Sheba *et al.* 1961	102(c)	100	2	1·96	
KERALA, Cochin		Sheba *et al.* 1961	58(c)	52	6	10·34	
MAHARASHTRA, Bombay†		Beaconsfield *et al.* 1967				14	
AFRICA							
ALGERIA and TUNISIA		Sheba *et al.* 1961	112(c)	111	1	0·89	
EGYPT		Sheba *et al.* 1961	112(c)	108	4	3·57	
ETHIOPIA†	Falasha	Adam 1962	225	225			
ETHIOPIA		Sheba *et al.* 1961	208	208			
LIBYA		Sheba *et al.* 1961	219(c)	217	2	0·91	
MOROCCO		Sheba *et al.* 1961	219(c)	218	1	0·46	
MOROCCO, Atlas		Sheba *et al.* 1961	23(c)	22	1	4·35	
TUNISIA, Île de Djerba		Sheba *et al.* 1961	52	52			
U.S.S.R.							
CAUCASUS		Sheba *et al.* 1961	25(c)	18	7	28·00	
UZBEKISTAN, Bukhara		Sheba *et al.* 1961	46	46			
TURKEY							
Urfa	Kurdish	Horowitz 1963	167	107	60	35·93	

TABLE 21.1.2
GLUCOSE-6-PHOSPHATE DEHYDROGENASE SYSTEM: NORMAL AND DEFICIENT TYPES; SEX NOT STATED

Place	Population	Authors	Number	Normal	Deficient	Deficient %
ISRAEL, Jaffa	Samaritans	Bonné 1966	132	132		
ISRAEL, Nablus	Samaritans	Bonné *et al.* 1967	37	37*		
IRAN		Ramot *et al.* 1962	62	53	9	14·52
KURDISTAN		Ramot *et al.* 1962	111	49	62	55·86
EGYPT	Karaites	Goldschmidt 1967	250	250		

*All B +

TABLE 21.1.3
GLUCOSE-6-PHOSPHATE DEHYDROGENASE SYSTEM: NORMAL AND DEFICIENT TYPES IN MALES AND FEMALES

Place	Population	Authors	Total	Males		Females	
				Number	Normal	Number	Normal
TURKEY, Istanbul†		Say *et al.* 1965	93	29	29	64	64

TABLE 21.1.4
GLUCOSE-6-PHOSPHATE DEHYDROGENASE SYSTEM: NORMAL AND DEFICIENT TYPES IN MALES; NORMAL, INTERMEDIATE, AND DEFICIENT TYPES IN FEMALES

Place	Population	Authors	Total	Males				Gd^- M. only	Females			Inter-mediate	Deficient	Gd^- M. & F.
				Number		Normal	Deficient		Number		Normal			
YEMEN, Habban		Bonné *et al.* 1970	572	284	Obs. Exp.	279 280·36	5 3·64	1·76	288	Obs. Exp.	282 280·66	6 7·29	0·05	1·28

TABLE 21.2
GLUCOSE-6-PHOSPHATE DEHYDROGENASE SYSTEM: NORMAL AND DEFICIENT TYPES; SUBTYPED IN MALES

Place	Population	Authors	Number	Normal		Deficient		Gd^{B+}	Gd^{B-}	Gd^{A+}
				A+	B+	A−	B−			
IRAQ, N.W.	Kurdish	Tills *et al.* 1977	36		11		25	30·56	69·44	
IRAQ, S.E.	Kurdish	Tills *et al.* 1977	19		8		11			
IRAN	Kurdish	Tills *et al.* 1977	63		40		23	63·49	36·51	
YEMEN, N. & N.E.		Tills *et al.* 1977	31		30		1	96·77	3·23	
YEMEN, S.		Tills *et al.* 1977	39		38		1	97·44	2·56	
YEMEN, Saada, San'a, Damar, Beida, Aden		Bodmer *et al.* 1972	71(a)	1	66		4	92·96	5·63	1·41

TABLE 22
6-PHOSPHOGLUCONATE DEHYDROGENASE SYSTEM

Place	Population	Authors	Number	Usual Type AA	Variants			PGD^A	PGD^C	$PGD^{Hackney}$
					Common AC	Canning CC	Hackney			
ISRAEL, Nablus	Samaritans	Bonné et al. 1967	37	28	7	2		85·14	14·86	
IRAQ, N.W.	Kurdish	Tills et al. 1977	61	56	5			95·90	4·10	
IRAQ, S.E.	Kurdish	Tills et al. 1977	50	47	3			97·00	3·00	
IRAN	Kurdish	Tills et al. 1977	106	96	9		1	95·28	4·25	0·47
YEMEN, N. & N.E.		Tills et al. 1977	23	18	5			89·13	10·87	
YEMEN, S.		Tills et al. 1977	88	78	9	1		93·75	6·25	
YEMEN, Habban		Bonné et al. 1970	551	481	66	4		93·28	6·72	
YEMEN, Habban		Fitch & Parr[1]	499	429	66	4		92·59	7·41	
YEMEN, Saada, San'a, Damar, Beida, Aden		Bodmer et al. 1972	201(a)	182	15	4		94·28	5·72	

[1] Quoted by Carter et al. 1968.

TABLE 23
PHOSPHOGLUCOMUTASE PGM$_1$ SYSTEM

Place	Population	Authors	Number	1	2-1	2	8-2	PGM_1^1	PGM_1^2	PGM_1^8	χ^2
ISRAEL	Ashkenazi	Szeinberg & Tomashevsky-Tamir 1971	185	120	53	12		79·19	20·81		3·17
ISRAEL, Nablus	Samaritans	Bonné et al. 1967	37	26	10	1		83·78	16·22		<0·01
IRAQ		Szeinberg & Tomashevsky-Tamir 1971	186	83	87	16		68·01	31·99		1·04
IRAQ, Hit	Karaites	Goldschmidt et al. 1976	70	36	27	7		70·71	29·29		0·33
IRAQ, N.W.	Kurdish	Tills et al. 1977	61	42	18	1		83·61	16·39		0·35
IRAQ, S.E.	Kurdish	Tills et al. 1977	50	24	22	4		70·00	30·00		0·11
IRAN	Kurdish	Tills et al. 1977	106	65	39	7		77·36	22·64		0·75
YEMEN		Szeinberg & Tomashevsky-Tamir 1971	192	100	80	12		72·92	27·08		0·58
YEMEN, N. & N.E.		Tills et al. 1977	65	22	31	12		57·69	42·31		0·03
YEMEN, S.		Tills et al. 1977	92	51	36	5		75·00	25·00		0·17
YEMEN, Habban		Bonné et al. 1970	297	59	134	104		42·42	57·58		1·73
YEMEN, Saada, San'a, Damar, Beida, Aden		Bodmer et al. 1972	202(a)	114	80	8		76·24	23·76		1·75
NORTHERN AFRICA		Szeinberg & Tomashevsky-Tamir 1971	183	89	76	17	1	69·40	30·33	0·27	2·43
LIBYA		Bonné-Tamir 1974	147*	93	46	8		78·91	21·09		0·52
GERMANY		Bonné-Tamir 1975	93	41	41	11		66·13	33·87		0·02

*No bands appeared in one more specimen.

TABLE 24
PHOSPHOGLUCOMUTASE PGM$_2$ SYSTEM

Place	Population	Authors	Number	Common	PGM_2^1
				1	
IRAQ	Sampled in England	Hopkinson & Harris 1966	69	69	100·00

TABLE 25
ADENYLATE KINASE SYSTEM

Place	Population	Authors	Number	AK1	AK2-1	AK2	AK^1	AK^2
ISRAEL	Ashkenazi	Szeinberg & Tomashevsky-Tamir 1971	191	170	21		94·50	5·50
ISRAEL, Nablus	Samaritans	Bonné *et al.* 1967	37	37			100·00	
IRAQ		Adam 1967	139	126	13		95·32	4·68
IRAQ		Szeinberg & Tomashevsky-Tamir 1971	190	179	11		97·11	2·89
IRAQ, Hit	Karaites	Goldschmidt *et al.* 1976	71	71			100·00	
IRAQ, N.W.	Kurdish	Tills *et al.* 1977	61	60	1		99·18	0·82
IRAQ, S.E.	Kurdish	Tills *et al.* 1977	50	48	2		98·00	2·00
IRAN	Kurdish	Tills *et al.* 1977	106	91	15		92·92	7·08
YEMEN		Szeinberg & Tomashevsky-Tamir 1971	193	179	14		96·37	3·63
YEMEN, N. & N.E.		Tills *et al.* 1977	65	59	6		95·38	4·62
YEMEN, S.		Tills *et al.* 1977	92	81	11		94·02	5·98
YEMEN, Saada, San'a, Damar, Beida, Aden		Bodmer *et al.* 1972	201(a)	187	14		96·52	3·48
NORTHERN AFRICA		Szeinberg & Tomashevsky-Tamir 1971	188	167	21		94·41	5·59
LIBYA		Bonné-Tamir 1974	148	141	7		97·64	2·36
GERMANY		Bonné-Tamir 1975	93	78	14	1	91·40	8·60

TABLE 26
LACTATE DEHYDROGENASE SYSTEM

Place	Population	Authors	Number	LDH 1-1
ISRAEL	Samaritans	Bonné *et al.* 1967	37	37
YEMEN, Saada, San'a, Damar, Beida, Aden		Bonné-Tamir 1977	199	199

TABLE 27
MALATE DEHYDROGENASE SYSTEM

Place	Population	Authors	Number	MDH 1-1
ISRAEL, Nablus	Samaritans	Bonné *et al.* 1967	37	37
ISRAEL	Samaritans	Leakey *et al.* 1972	39	39
KURDISTAN		Leakey *et al.* 1972	99	99
YEMEN		Leakey *et al.* 1972	204	204
YEMEN, Habban		Leakey *et al.* 1972	597	597

TABLE 28
PEPTIDASE A SYSTEM

Place	Population	Authors	Number	PepA 1	PepA 2-1	PepA 2	$PepA^1$	$PepA^2$
GERMANY		Bonné-Tamir 1975	93	92	1		99·46	0·54

TABLE 29
ADENOSINE DEAMINASE SYSTEM

Place	Population	Authors	Number	ADA 1	ADA 2-1	ADA 2	ADA^1	ADA^2
ISRAEL	Ashkenazi	Szeinberg et al. 1971	437	348	85	4	89·36	10·64
IRAQ		Szeinberg et al. 1971	291	211	73	7	85·05	14·95
IRAQ, N.W.	Kurdish	Tills et al. 1977	41	22	18	1	75·61	24·39
IRAQ, S.E.	Kurdish	Tills et al. 1977	41	33	8		90·24	9·76
IRAN	Kurdish	Tills et al. 1977	19	11	8			
YEMEN		Szeinberg et al. 1971	219	164	51	4	86·53	13·47
YEMEN, N. & N.E.		Tills et al. 1977	43	35	8		90·70	9·30
YEMEN, S.		Tills et al. 1977	41	34	7		91·46	8·54
YEMEN, Saada, San'a, Damar, Beida, Aden		Bonné-Tamir 1977	170	121	45	4	84·41	15·59
NORTHERN AFRICA		Szeinberg et al. 1971	204	168	35	1	90·93	9·07
LIBYA		Bonné-Tamir 1974	148	135	13		95·61	4·39
GERMANY		Bonné-Tamir 1975	93	75	17	1	89·78	10·22

TABLE 30
GLUTAMIC-PYRUVIC TRANSAMINASE SYSTEM

Place	Population	Authors	Number	Gpt 1-1	Gpt 2-1	Gpt 2-2	Gpt^1	Gpt^2	χ^2
ISRAEL	Ashkenazi	Lahav & Szeinberg 1972	196	70	95	31	59·95	40·05	0·02
IRAQ		Lahav & Szeinberg 1972	192	66	94	32	58·85	41·15	0·02
YEMEN		Lahav & Szeinberg 1972	190	99	78	13	72·63	27·37	0·18
NORTHERN AFRICA		Lahav & Szeinberg 1972	193	85	89	19	67·10	32·90	0·38

TABLE 31
PHENYLTHIOCARBAMIDE TASTING SYSTEM

Place	Population	Authors	Number	Tasters	Non-Tasters	T	t	
SOUTH-WEST ASIA								
ISRAEL	Ashkenazi	Younovitch 1934[1]	245(c)	168	77	43·94	56·06	Filter paper
ISRAEL	Ashkenazi	Brand 1963	674	642	32	78·21	21·79	Threshold
ISRAEL, Jaffa	Samaritans	Bonné 1966	125	117	8	74·70	25·30	Threshold
ISRAEL	Sephardi	Younovitch 1934[1]	175(c)	126	49	47·08	52·92	Filter paper
IRAQ, Baghdad		Boyd & Boyd 1941	168	139	29	58·45	41·55	Filter paper
IRAQ and IRAN		Sheba et al. 1962	336	282	54	59·91	40·09	Threshold
KURDISTAN		Sheba et al. 1962	129	111	18	62·65	37·35	Threshold
KURDISTAN		Guttman et al. 1967	455(c)	333	122	48·22	51·78	Threshold
YEMEN		Younovitch 1934[1]	59(c)	40	19	43·25	56·75	Filter paper
YEMEN		Sheba et al. 1962	261	214	47	57·56	42·44	Threshold
YEMEN		Guttman et al. 1967	498(c)	367	131	48·71	51·29	Threshold
INDIA								
BOMBAY†	Bene Israel	Sirsat 1956	200(c)	160	40	55·28	44·72	Single solution
BOMBAY†	Baghdadi	Sirsat 1956	200(c)	159	41	54·72	45·28	Single solution
KERALA, Cochin		Sheba et al. 1962	41	28	13	43·69	56·31	Threshold
KERALA, Cochin		Guttman et al. 1967	402(c)	233	169	35·16	64·84	Threshold
AFRICA								
LIBYA		Guttman et al. 1967	501(c)	389	112	52·71	47·29	Threshold
MOROCCO, TRIPOLITANIA, TUNISIA		Sheba et al. 1962	340	289	51	61·27	38·73	Threshold
MOROCCO, South Atlas		Guttman et al. 1967	464(c)	308	156	42·02	57·98	Threshold
TUNISIA, île de Djerba		Sheba et al. 1962	41	24	17	35·61	64·39	Threshold
TUNISIA, île de Djerba		Guttman et al. 1967	383(c)	222	161	35·16	64·84	Threshold
ETHIOPIA†	Falasha	Bat-Miriam et al. 1962	133(c)	122	11	71·24	28·76	Threshold
EUROPE								
CZECHOSLOVAKIA, HUNGARY, ROMANIA	Ashkenazi	Sheba et al. 1962	121	96	25	54·55	45·45	Threshold
LITHUANIA, POLAND, W. RUSSIA	Askhenazi	Sheba et al. 1962	319	253	66	54·51	45·49	Threshold
POLAND†		Modrzewska 1958	98	73	25	49·49	50·51	Filter paper
BALKAN COUNTRIES	Sephardi	Sheba et al. 1962	101	79	22	53·33	46·67	Threshold
AMERICA								
U.S.A., California, Los Angeles†	Of Central European and Russian origin	Terry & Segall 1947	138	106	32	51·84	48·16	Filter paper
U.S.A., Ohio†		Rife & Schonfeld 1944	82	70	12	61·75	38·25	Filter paper
BRAZIL, São Paulo†	Ashkenazi	Saldanha & Beçak 1959	244	176	68	47·21	52·79	Threshold

[1] Quoted by Parr 1934.

TABLE 32
ACETYLATOR SYSTEM: RAPID AND SLOW TYPES

Place	Population	Authors	Number	Rapid	Slow	Ac^R	Ac^S
ISRAEL	Ashkenazi	Szeinberg et al. 1961	100	37	63	20·63	79·37
ISRAEL	Oriental and Sephardi	Szeinberg et al. 1961	179	65	114	20·19	79·81
U.S.A.†		Harris, H. W. 1961	11	5	6		

TABLE 33
PRIMARY ADULT LACTASE DEFICIENCY—LACTOSE TOLERANCE TESTS

Place	Population	Authors	Number tested	Lactose intolerant* Number	Per cent
LEBANON, SYRIA, IRAN	Oriental	Gilat et al. 1970	34(a)	29	85·29
IRAQ	Oriental	Gilat et al. 1970	43(a)	33	76·74
YEMEN	Oriental	Gilat et al. 1970	46(a)	25	54·35
LIBYA, MOROCCO, TUNISIA	Sephardi	Gilat et al. 1970	41(a)	22	53·66
BULGARIA, GREECE, TURKEY	Sephardi	Gilat et al. 1970	81(a)	57	70·37
CZECHOSLOVAKIA, GERMANY, HUNGARY, LITHUANIA, POLAND, ROMANIA, RUSSIA	Ashkenazi	Gilat et al. 1970	135(a)	87	64·44

Subjects include family members.
*Plasma glucose rise less than 20 mg/100 ml.

BIBLIOGRAPHY

ABRAMSKY, C. (1974). An opportunity missed. *Soviet Jewish Affairs* **4**, 95–100. [Review of *The Jews of Poland* by B. D. Weinryb. Jewish Publ. Soc. of America, Philadelphia, 1973.]

ABRAMSKY, C. (1976). The Khazar myth. *Jewish Chron.* Apr. 6, 19. [Review of A. Koestler 1976.]

ADAIR, H. & OWENS, J. (1962). Pemphigus vulgaris: a study of 34 cases at Charity Hospital, New Orleans. *Sth med. J.* **55**, 1034–9.

ADAM, A. (1962). A survey of some genetical characters in Ethiopian tribes. i. Glutathione stability and glucose-6-phosphate dehydrogenase activity in red blood cells. *Amer. J. phys. Anthrop.* **20**, 172–3.

ADAM, A. (1967). Quoted by RAPLEY, S. *et al.* (1967).

ADAM, A., TIPPETT, P., GAVIN, J., NOADES, J., SANGER, R., & RACE, R. R. (1967). The linkage relation of Xg to g-6-pd in Israelis: the evidence of a second series of families. *Ann. hum. Genet.* **30**, 211–18.

ADAM, A. (1973). Genetic diseases among Jews. *Israel J. med. Sci.* **9**, 1383–92.

ALBRIGHT, W. F. (1960). *The archaeology of Palestine.* Penguin Books, Harmondsworth.

ALLEGRO, J. M. (1961). *The Dead Sea Scrolls.* Penguin Books, Harmondsworth.

ALLEN, F. H. & LEWIS, S. J. (1957). Kpª (Penny), a new antigen in the Kell blood group system. *Vox sang.* n.s. **2**, 81–7.

ALLEN, F. H., LEWIS, S. J., & FUDENBERG, H. (1958). Studies of anti-Kpᵇ, a new antibody in the Kell blood group system. *Vox sang.* n.s. **3**, 1–13.

ALLISON, A. C. (1954). The distribution of the sickle-cell trait in East Africa and elsewhere, and its apparent relationship to subtertian malaria. *Trans. R. Soc. trop. Med. Hyg.* **48**, 312–18.

ALLISON, A. C. (1954). Protection afforded by sickle-cell trait against subtertian malarial infection. *Brit. med. J.* **i**, 290–4.

ALLISON, A. C. & BLUMBERG, B. S. (1961). An isoprecipitation reaction distinguishing human serum-protein types. *Lancet* **i**, 634–7.

ALLISON, A. C. & CLYDE, D. F. (1961). Malaria in African children with deficient erythrocyte glucose-6-phosphate dehydrogenase. *Brit. med. J.* **i**, 1346–9.

ALTER, A., GELB, A. G., CHOWN, B., ROSENFIELD, R. E., & CLEGHORN, T. E. (1967). Gonzales (Goª), a new blood group character. *Transfusion, Philad.* **7**, 88–91.

ALTER, M. (1974). Creutzfeldt–Jacob disease: hypothesis for high incidence in Libyan Jews in Israel. *Science* **186**, 848.

ALTOUNYAN, E. H. R. (1928). Blood group percentages for Arabs, Armenians and Jews. Analysis of 1·758 groupings. *Brit. med. J.* **1**, 546.

AUZAS, C. (1957). Pers. comm. to A. E. Mourant.

AVDEYEVA & GRYTSEVICH. Quoted by HIRSZFELD, L. (1928).

BAER, Y. (1961). *A history of the Jews in Christian Spain.* Jewish Publ. Soc. of America, Philadelphia.

BARINSTEIN, L. A. (1928). Zur Frage des biochemischen Rassenindex der Bevölkerung von Odessa. *Ukr. Zbl. Blutgr.* **2**, 58–61.

BAR-SHANY, S. (1974). Pers. comm. to A. E. Mourant.

BAT-MIRIAM, M. (1962). A survey of some genetical characters in Ethiopian tribes. iv. The blood groups of the Falasha, Galla and Guraghe tribes. *Amer. J. phys. Anthrop.* **20**, 179–82.

BAT-MIRIAM, M., ADAM, A., & HANANEL, Z. (1962). A survey of some genetical characters in Ethiopian tribes. vi. Taste thresholds for phenylthiourea. *Amer. J. phys. Anthrop.* **20**, 190–3.

BAUM, G. L., RACZ, I., BUBIS, J. J., MOLCHO, M., & SHAPIRO, B. L. (1966). Cystic disease of the lung. Report of eighty-eight cases, with an ethnologic relationship. *Amer. J. Med.* **40**, 578–602.

BEACONSFIELD, P., MAHBOUBI, E., & RAINSBURY, R. (1967). Epidemiologie des Glukose-6-Phosphat-Dehydrogenase-Mangels. *Münch. med. Wschr.* **109**, 1950–2.

BEAVEN, G. H. (1973). Biological studies of Yemenite and Kurdish Jews in Israel and other groups in south-west Asia. X. Haemoglobin studies of Yemenite and Kurdish Jews in Israel. *Phil. Trans.* B, **266**, 185–93.

BEHZAD, O., LEE, C. L., GAVIN, J., & MARSH, W. L. (1973). A new anti-erythrocyte antibody in the Duffy system: anti-Fy4. *Vox Sang.* **24**, 337–42.

BEN-ZVI, I. (1958). *The exiled and the redeemed. The strange Jewish 'tribes' of the Orient.* Vallentine, Mitchell, London.

BIAS, W. B., LIGHT-ORR, J. K., KREVANS, J. R., HUMPHREY, R. L., HAMILL, P. V. V., COHEN, B. H., & McKUSICK, V. A. (1969). The Stoltzfus blood group, a new polymorphism in man. *Amer. J. hum. Genet.* **21**, 552–8.

BIENZLE, U., AYENI, O., LUCAS, A. O., & LUZZATTO, L. (1972). Glucose-6-phosphate dehydrogenase and malaria. Greater resistance of females heterozygous for enzyme deficiency and of males with non-deficient variant. *Lancet* **i**, 107–10.

BIJLMER, H. J. T. (1943). Bloedgroepenonderzoek bij de bevolking van Amsterdam. *Ned. Tijdschr. Geneesk.* **87**, 1467–70.

BJARNASON, O., BJARNASON, V., EDWARDS, J. H., FRIDRIKSSON, S., GODBER, M. J., MOURANT, A. E., TILLS, D., & WOODHEAD, B. G. (1968). Unpubl. observations.

BLOOM, D. (1966). The syndrome of congenital telangiectatic erythema and stunted growth. *J. Pediat.* **68**, 103–13.

BLUMBERG, B. S. (1963). Polymorphisms of the human serum proteins and other biological systems. In *The genetics of migrant and isolate populations*, (ed. E. Goldschmidt), pp. 20–6. Williams and Wilkins, Baltimore.

BODMER, J., BONNÉ, B., BODMER, W., BLACK, S., BEN-DAVID, A. & ASHBEL, S. (1972). Study of the HL-A system in a Yemenite Jewish population in Israel. In *Histocompatibility testing 1972*, (ed. J. Dausset & J. Colombani), pp. 125–32. Munksgaard, Copenhagen [1973].

BONNÉ, B. (1963). The Samaritans: a demographic study. *Hum. Biol.* **35**, 61–89.

BONNÉ, B. (1966). Genes and phenotypes in the Samaritan isolate. *Amer. J. phys. Anthrop.* **24**, 1–19.

BONNÉ, B. *et al.* (1967). Unpubl. observations.

BONNÉ, B., ASHBEL, S., MODAI, M., GODBER, M. J., MOURANT, A. E., TILLS, D., & WOODHEAD, B. G. (1970). The Habbanite isolate. i. Genetic markers in the blood. *Hum. Hered.* **20**, 609–22.

BONNÉ, B., ASHBEL, S., & TAL, A. (1971). The Habbanite isolate. ii. Digital and palmar dermatoglyphics. *Hum. Hered.* **21**, 478–92.

BONNÉ, B., ASHBEL, S., BERLIN, G., & SELA, B. (1972). The Habbanite isolate. iii. Anthropometrics, taste sensitivity and color vision. *Hum. Hered.* **22**, 430–44.

BONNÉ-TAMIR, B. (1974–77). Pers. comm. to A. E. Mourant.

BOTTINI, E., LUCARELLI, P., AGOSTINO, R., PALMARINO, R., BUSINCO, L., & ANTOGNONI, G. (1971). Favism: association with erythrocyte acid phosphatase phenotype. *Science* **171**, 409–11.

BOYD, W. C. & BOYD, L. G. (1937). New data on blood groups and other inherited factors in Europe and Egypt. *Amer. J. phys. Anthrop.* **23**, 49–70.

BOYD, W. C. (1939). Blood groups. *Tabul. Biol., Hague* **17**, 113–240.

BOYD, W. C. & BOYD, L. G. (1941). Blood groups and types in Baghdad and vicinity. *Hum. Biol.* **13**, 398–404.

BRAND, N. (1963). Taste sensitivity and endemic goitre in Israel. *Ann. hum. Genet.* **26**, 321–4.

BRANFOOT, A. C. (1971). Consumption of sheep's brains by patients with multiple sclerosis. *Lancet* **i**, 1235.

BRIGGS, L. C. (1958). *The living races of the Sahara Desert.* Peabody Museum, Cambridge, Mass.

BRIGGS, L. C. & GUÈDE, N. L. (1964). *No more for ever. A Saharan Jewish town.* The Peabody Museum, Cambridge, Mass.

BRONTE-STEWART, B., BOTHA, M. C., & KRUT, L. H. (1962). ABO blood groups in relation to ischaemic heart disease. *Brit. med. J.* **i**, 1646–50.

BRZEZINSKI, A., GUREVITCH, J., HERMONI, D., & MUNDEL, G. (1952). Blood groups in Jews from the Yemen. *Ann. Eug.* **16**, 335–7.

BUNAK, V. V. Quoted by Wagner, L. B. (1926).

CAILLON, L. & DISDIER, C. (1930). A propos des groupes sanguins, leurs rapports avec les différentes races de la Tunisie. *Arch. Inst. Pasteur Tunis.* **19**, 41–9.

CARTER, N. D., FILDES, R. A., FITCH, L. I., & PARR, C. W. (1968). Genetically determined electrophoretic variations of human phosphogluconate dehydrogenase. *Acta genet. Stat. med.* **18**, 109–22.

CAVALLI-SFORZA, L. L. & CARMELLI, D. (1977). Pers. comm. to A. E. Mourant.

CHEN, S. H. & GIBLETT, E. R. (1971). Polymorphism of soluble glutamic-pyruvic transaminase: a new genetic marker in man. *Science* **173**, 148–9.

CHEN, S. H., GIBLETT, E. R., ANDERSON, J. E., & FOSSUM, B. L. G. (1972). Genetics of glutamic-pyruvic transaminase: its inheritance, common and rare variants, population distribution, and differences in catalytic activity. *Ann. hum. Genet.* **35**, 401–9.

CHOWN, B., PETERSON, R. F., LEWIS, M., & HALL, A. (1949). On the ABO gene and Rh chromosome distribution in the white population of Manitoba. *Can. J. Res.*, E, **27**, 214–25.

CHOWN, B. & LEWIS, M. (1951). Pers. comm. to A. E. Mourant.

CLEVE, H., RAMOT, B., & BEARN, A. G. (1962). Distribution of the serum group specific components in Israel. *Nature, Lond.* **195**, 86–7.

COGGINS, R. J. (1975). *Samaritans and Jews.* Basil Blackwell, Oxford.

COHEN, A. I. (196?, date unknown). *Mount Gerizim.* Greek Convent Press, Jerusalem.

COHEN, T. (1971). Genetic markers in migrants to Israel. *Israel J. med. Sci.* **7**, 1509–14.

COHEN, T., BRAND-AURABAN, A., KARSHAI, C., JACOB, A., GAY, I., TSITSIANOV, J., SHAPIRO, T., JATZIN, S., & ASHKENAZI, A. (1973). Familial infantile renal tubular acidosis and congenital nerve deafness: an autosomal recessive syndrome. *Clin. Genet.* **4**, 275–8.

COOMBS, R. R. A., MOURANT, A. E., & RACE, R. R. (1946). *In-vivo* isosensitization of red cells in babies with haemolytic disease. *Lancet* **i**, 264–6.

CUTBUSH, M., MOLLISON, P. L., & PARKIN, D. M. (1950). A new human blood group. *Nature, Lond.* **165**, 188.

DARÁNYI, J. VON. (1940). Die Blutgruppen in Ungarn, insbesondere die der führenden Volksschicht. *Z. Immun-Forsch.* **99**, 77–85.

DAUSSET, J. & COLOMBANI, J., (eds.) (1973). *Histocompatibility testing 1972.* Munksgaard, Copenhagen.

DAUSSET, J. & SVEJGAARD, A., (eds.) (1977). *HLA and disease.* Munksgaard, Copenhagen and Williams & Wilkins, Baltimore.

DONEGANI, J. A., IBRAHIM, K., IKIN, E. W., & MOURANT, A. E. (1950). The blood groups of the people of Egypt. *Heredity* **4**, 377–82.

DRESSLER, L. (1951). The blood group distribution among the Jews in Israel. *4th Int. Congr. Blood Transfus.*, Lisbon, 1951, 388–9.

DREYFUSS, F., IKIN, E. W., LEHMANN, H., & MOURANT, A. E. (1952). An investigation of blood-groups and a search for sickle-cell trait in Yemenite Jews. *Lancet* **ii**, 1010–12.

DREYFUSS, F., MUNDEL, G., & BENYESCH, M. (1953). Sickling in Oriental Jews. *Acta haemat.* **9**, 193–9.

DUBIN, I. N. & JOHNSON, F. B. (1954). Chronic idiopathic jaundice with unidentified pigment in liver cells. *Medicine* **33**, 155–97.

DUNN, C. (1955). A genetical study of the nuclear (ghetto) Jewish community of Rome. *Amer. J. phys. Anthrop.* **13**, 389 (summary).

DUNN, L. C. & DUNN, S. P. (1957). The Jewish community of Rome. *Sci. Amer.* **196**, 119–28.

EISENBERG, A. (1928). Zur Frage nach den Isoagglutinationsgruppen des Blutes bei Menschen. *Folia haemat., Lpz.* **36**, 316–36.

EPSTEIN, I. (1959). *Judaism. A historical presentation.* Penguin Books, Harmondsworth.

EVANS, D. A. P., MANLEY, K. A., & MCKUSICK, V. A. (1960). Genetic control of isoniazid metabolism in man. *Brit. med. J.* **ii**, 485–91.

FELDMAN & ELMANOVITCH. Quoted by HIRSZFELD, L. (1928).

FEMARO, J. & ALKAN, W. J. (1968). Familial neutropenia in Jews of Yemenite origin. *9th Int. Congr. Life Assurance Med.*, 172–7.

FERGUSON, A. & MAXWELL, J. D. (1967). Genetic aetiology of lactose intolerance. *Lancet* **ii**, 188–90.

FINK, D. (1971). Consumption of sheep's brains by patients with multiple sclerosis. *Lancet* **ii**, 1235.

FISHBERG, M. (1911). *The Jews: a study of race and environment.* Walter Scott Publishing Co., London & Felling-on-Tyne.

FORBAT, A., LEHMANN, H., & SILK, E. (1953). Prolonged apnoea following injection of succinyldicholine. *Lancet* ii, 1067.

FOX, A. L. (1931). Six in ten 'tasteblind' to bitter chemical. *Sci. News Lett., Wash.* 19, 249.

FRIED, K., BLOCH, N., SUTTON, E., NEEL, J. V., BAYANI-SIOSON, P., RAMOT, B., & DUVDEVANI, P. (1963). Haptoglobins and transferrins. In *The genetics of migrant and isolate populations*, (ed. Elisabeth Goldschmidt), pp. 266–7. Williams and Wilkins, Baltimore.

FUHRMANN, W. (1967). Genetic aspects of lipidoses. In *Lipids and lipidoses* (ed. G. Schettler), pp. 493–525. Springer, New York.

FURUHJELM, U., NEVANLINNA, H. R., NURKKA, R., GAVIN, J., TIPPETT, P., GOOCH, A., & SANGER, R. (1968). The blood group antigen Ul^a (Karhula). *Vox Sang.* 15, 118–24.

GAMST, F. C. (1969). *The Qemant. A pagan-Hebraic peasantry of Ethiopia*. Holt, Rinehart and Winston, New York.

GASTER, M. (1925). *The Samaritans: their history, doctrines and literature*. Oxford University Press.

GAUD, J. & MÉDIONI, L. (1948). Sur la répartition des groupes sanguins au Maroc. *Bull. Inst. Hyg. Maroc.* 8 n.s., 91–101.

GEKKER & GOROCHZAN. Quoted by SEMENSKAYA, E. M. (1930).

GELPI, A. P. (1965). Glucose-6-phosphate dehydrogenase deficiency in Saudi Arabia: a survey. *Blood.* 25, 486–93.

GELPI, A. P. (1967). Glucose-6-phosphate dehydrogenase deficiency, the sickling trait, and malaria in Saudi Arab children. *J. Pediat.* 71, 138–46.

GENNA, G. E. (1938). *I Samaritani*. Roma, Publ. Com. ital. Stud. Popol., ser. 5, Spedizioni scientifiche, v. 1.

GIBLETT, E. R. (1958). Js, a 'new' blood group antigen found in Negroes. *Nature, Lond.* 181, 1221–2.

GIBLETT, E. R. (1969). *Genetic markers in human blood*. Blackwell Scientific Publications, Oxford.

GILAT, T., KUHN, R., GELMAN, E., & MIZRAHY, O. (1970). Lactase deficiency in Jewish communities in Israel. *Amer. J. digest. Dis.* 15, 895–904.

GILAT, T., GELMAN-MALACHI, E., & SHOCHET, S. B. (1971). Lactose tolerance in an Arab population. *Amer. J. digest. Dis.* 16, 202–3.

GILAT, T., BENAROYA, Y., GELMAN-MALACHI, E., & ADAM, A. (1973). Genetics of primary adult lactase deficiency. *Gastroenterology* 64, 562–8.

GODBER, M. J., KOPEĆ, A. C., MOURANT, A. E., TILLS, D., & LEHMANN, E. E. (1973). Biological studies of Yemenite and Kurdish Jews in Israel and other population groups in south-west Asia. ix. The hereditary blood factors of the Yemenite and Kurdish Jews. *Phil. Trans.* B, 266, 169–84.

GOLDSCHMIDT, E. (1967). Summary and conclusions. *9th Int. Congr. Life Ass. Med.*, Tel-Aviv, 200–6.

GOLDSCHMIDT, E., COHEN, T., ISACSOHN, M., & FREIER, S. (1968). Incidence of hemoglobin Bart's in a sample of newborn from Israel. *Acta genet. Stat. med.* 18, 361–8.

GOLDSCHMIDT, E., FRIED, K., STEINBERG, A. G., & COHEN, T. (1976). The Karaite community of Iraq in Israel: a genetic study. *Amer. J. hum. Genet.* 28, 243–52.

GRIGOROVA, O. (1931). Die Isoagglutination bei Kindern im Zusammenhang mit den Konstitutionellen Eigenschaften. *Z. Rassenphysiol.* 4, 155–63.

GROEN, J. J. (1964). Gaucher's disease: hereditary transmission and racial distribution. *Arch. intern. Med.* 113, 543–9.

GRUBINA, A. W. (1930). Die Blutgruppen unter den Schul-kindern der Bevölkerung von Ksyl-Orda (Kasakstan). *Ukr. Zbl. Blutgr.* 4, 240–5.

GRÜNEBERG, H. (1936). Two independent inherited tooth anomalies in one family. *J. Hered.* 27, 225–8.

GUNSON, H. H. & LATHAM, V. (1972). An agglutinin in human serum reacting with cells from Le(a-b-) non-secretor individuals. *Vox Sang.* 22, 344–53.

GUREVITCH, J. (1940). *Blood group and blood transfusion*. Jerusalem, Hadassah Univ. Hosp. (In Hebrew).

GUREVITCH, J., BRZEZINSKI, A., & POLISHUK, Z. (1947). Rh factor and haemolytic disease of the newborn in Jerusalem Jews. *Lancet* ii, 943.

GUREVITCH, J., HERMONI, D., & POLISHUK, Z. (1951). *Rh* blood types in Jerusalem Jews. *Ann. Eugen., Lond.* 16, 129–30.

GUREVITCH, J., HERMONI, D., & MARGOLIS, E. (1953). Blood groups in Kurdistani Jews. *Ann. Eugen., Lond.* 18, 94–5.

GUREVITCH, J., HASSON, E., MARGOLIS, E., & POLIAKOFF, C. (1954). Blood groups in Jews from Tripolitania and Cochin, India. *5th Int. Congr. Blood Transfus*, Paris, 250–3.

GUREVITCH, J. & MARGOLIS, E. (1954–5). Blood groups in Jews from Iraq. *Ann. hum. Genet.* 19, 257–9.

GUREVITCH, J., HASSON, E., & MARGOLIS, E. (1956). Blood groups in Persian Jews. A comparative study with other oriental Jewish communities. *Ann. hum. Genet.* 21, 135–8.

GUREVITCH, J. (1965). Quoted by PALATNIK, M. (1965).

GUTTMAN, R., GUTTMAN, L., & ROSENZWEIG, K. A. (1967). Cross-ethnic variation in dental, sensory and perceptual traits: a nonmetric multibivariate derivation of distances for ethnic groups and traits. *Amer. J. phys. Anthrop.* 27, 259–76.

HAEMMERLI, U. P., KISTLER, H., AMMANN, R., MARTHALER, T., SEMENZA, G., AURICCHIO, S., & PRADER, A. (1965). Acquired milk intolerance in the adult caused by lactose malabsorption due to a selective deficiency of intestinal lactase deficiency. *Amer. J. Med.* 38, 7–30.

HALBER, W. & MYDLARSKI, J. (1925). Untersuchungen über die Blutgruppen in Polen. *Z. ImmunForsch.* 43, 470–84.

HALDANE, J. B. S. (1949). Mutation in man. *Hereditas, Lund.* suppl., 267–73.

HARRIS, H., KALMUS, H., & TROTTER, W. R. (1949). Taste sensitivity to phenylthiourea in goitre and diabetes. *Lancet* ii, 1038–9.

HARRIS, H. (1970). *The principles of human biochemical genetics*. North-Holland, Amsterdam and London.

HARRIS, H. W., KNIGHT, R. A., & SELIN, M. J. (1958). Comparison of isoniazid concentrations in the blood of people of Japanese and European descent. *Amer. Rev. Tuberc.* 78, 944–8.

HARRIS, H. W. (1961). Isoniazid metabolism in humans: genetic control, variation among races and influence on the chemotherapy of tuberculosis. *16th Int. Tuberc. Conf.*, Toronto 2, 503–7.

HARRIS, R. & GILLES, H. M. (1961). Glucose-6-phosphate dehydrogenase deficiency in the peoples of the Niger Delta. *Ann. hum. Genet.* 25, 199–206.

HARRISON, G. A., KÜCHEMANN, C. F., MOORE, M. A. S., BOYCE, A. J., BAJU, T., MOURANT, A. E., GODBER, M. J., GLASGOW, B. G., KOPEĆ, A. C., TILLS, D., & CLEGG, E. J. (1969). The effects of altitudinal variation in Ethiopian populations. *Phil. Trans.* B. 256, 147–82.

HEATON, E. W. (1968). *The Hebrew Kingdoms*. Oxford University Press.

HEDAYAT, S., AMIRSHANY, P., & KHADEMY, B. (1969). Frequency of G-6-PD deficiency among some Iranian ethnic groups. *Trop. geogr. Med.* **21**, 163–8.

HELLER, H., SOHAR, E., & SHERF, L. (1958). Familial Mediterranean fever. *Arch. intern. Med.* **102**, 50–71.

HERWERDEN, M. A. VAN & BOELE-NIJLAND, T. Y. (1930). Investigation of blood groups in Holland. I. *Proc. Acad. Sci. Amst.* **33**, 659–73.

HERWERDEN, M. A. VAN. Quoted by STEFFAN, P. & WELLISCH, S. (1932).

HERZBERG, L., HERZBERG, B. N., GIBBS, C. J., SULLIVAN, W., AMYX, H., & GAJDUSEK, D. C. (1974). Creutzfeldt-Jacob disease: hypothesis for high incidence in Libyan Jews in Israel. *Science* **186**, 848.

HERZBERGER (1930). Quoted by BIJLMER, H. J. T. (1943).

HINDS, J. R. (1958). Bronchiectasis in the Maori. *N.Z. med. J.* **57**, 328–32.

HIRSCHFELD, J. (1959). Immune-electrophoretic demonstration of qualitative differences in human sera and their relation to the haptoglobins. *Acta path. microbiol. scand.* **47**, 160–8.

HIRSZFELD, L. & HIRSZFELD, H. (1918–19). Essai d'application des méthodes sérologiques au problème des races. *Anthropologie, Paris.* **29**, 505–37.

HIRSZFELD, L. (1928). *Konstitutionsserologie und Blutgruppenforschung*. Berlin. Translation in: US Army Med. Res. Lab. *Selected contributions to the literature of blood groups and immunology*. Fort Knox, Kentucky, 1969, **3**, pt. 1.

HITTI, P. K. (1961). *The Near East in history. A 5000 year story*. D. Van Nostrand Company, Inc., Princeton, N.J.

HOLMAN, C. A. (1953). A new rare human blood group-antigen (Wr^a). *Lancet* **ii**, 119–20.

HOOPER, I. (1947). Blood group distributions in Ireland. *Irish J. med. Sci.* July, 1–9 in reprint.

HOPKINSON, D. A., SPENCER, N., & HARRIS, H. (1963). Red cell acid phosphatase variants: a new human polymorphism. *Nature, Lond.* **199**, 969–71.

HOPKINSON, D. A. & HARRIS, H. (1966). Rare phosphoglucomutase phenotypes. *Ann. hum. Genet.* **30**, 167–81.

HOROWITZ, A. (1963). Polymorphic characters and genetic affinities of the Jewish community from Urfa. *Proc. Genet. Soc. Israel.* **12**, 219–21.

HUGHES, A. J. B., LOWE, R. F., GADD, K. G., & ELLIS, B. P. B. (1976). The sero-anthropology of the Rhodesian Lemba. Pers. comm. to A. E. Mourant.

IKIN, E. W., MOURANT, A. E., PETTENKOFER, H. J., & BLUMENTHAL, G. (1951). Discovery of the expected haemagglutinin anti-Fy^b. *Nature, Lond.* **168**, 1077.

IKIN, E. W. (1963). *The incidence of the blood antigens in different populations*. Ph.D. (Path.) Thesis, London.

IKIN, E. W., SMITH, H. M., BROOKS, P., & MOURANT, A. E. (1963). Unpubl. observations.

IKIN, E. W., MOURANT, A. E., KOPEĆ, A. C., LEHMANN, H., SCOTT, R. A. P., & HORSFALL, J. (1972). The blood groups and haemoglobin of the Jews of the Tafilalet oases of Morocco. *Man.* n.s. **7**, 595–600.

IONESCU, P. & IONESCU, E. (1930). Beiträge zum Studium der Blutgruppen in Rumänien. *Folia haemat., Lpz.* **42**, 91–8.

ISRAEL. MINISTRY OF LABOUR, SURVEY OF ISRAEL. (1970). *Atlas of Israel. Cartography, physical geography, human and economic geography, history*. Elsevier, Amsterdam.

ISSELBACHER, K. J., SCHEIG, R., PLOTKIN, G. R., & CAUL-

FIELD, J. B. (1964). Congenital β-lipoprotein deficiency: an hereditary disorder involving a defect in the absorption and transport of lipids. *Medicine* **43**, 347–61.

JETTMAR, H. M. (1930). Blutgruppenuntersuchungen in der Nordmongolei und der Nordmandschurei. *Mitt. anthrop. Ges. Wien.* **60**, 39–47.

JOSEPH, H. L. (1964). Pachonychia congenita. *Arch. Derm.* **90**, 594–603.

JOSHUA, H., SPITZER, A., & PRESENTEY, B. (1970). The incidence of peroxidase and phospholipid deficiency in eosinophilic granulocytes among various groups in Israel. *Amer. J. hum. Genet.* **22**, 574–7.

KAHANA, E., ALTER, M., BRAHAM, J., & SOFER, D. (1974). Creutzfeldt-Jacob disease: focus among Libyan Jews in Israel. *Science* **83**, 90–1.

KATTAMIS, C., ZANNOS-MARIOLEA, L., FRANCO, A. P., LIDDELL, J., LEHMANN, H., & DAVIES, D. (1962). Frequency of atypical pseudo-cholinesterase in British and Mediterranean populations. *Nature, Lond.* **196**, 599–600.

KATZ, S. I., DAHL, M. V., PENNEYS, N., TRAPANI, R. J., & ROGENTINE, N. (1973). HL-A antigens in pemphigus. *Arch. Derm.* **108**, 53–5.

KENNEDY, W. P. & MACFARLANE, J. (1936). Blood groups in Iraq. *Amer. J. phys. Anthrop.* **21**, 87–9.

KEVORKOV MARTIUKOV. Quoted by GUREVITCH, J. (1940).

KHACHADURIAN, A. K. (1962). Essential pentosuria. *Amer. J. hum. Genet.* **14**, 249–55.

KITCHIN, F. D., HOWEL-EVANS, W., CLARKE, C. A., McCONNELL, R. B., & SHEPPARD, P. M. (1959). P.T.C. taste response and thyroid disease. *Brit. med. J.* **i**, 1069–74.

KITCHIN, F. D. & BEARN, A. G. (1964). Distribution of serum group-specific components (Gc) in Afghanistan, Korean, Nigerian and Israeli populations. *Nature, Lond.* **202**, 827–8.

KOESTLER, A. (1976). *The Thirteenth Tribe. The Khazar Empire and its heritage*. Hutchinson, London.

KONUGRES, A. A., FITZGERALD, H., & DRESSER, R. (1965). Distribution and development of the blood factor Mt^a. *Vox Sang.* **10**, 206–7.

KOSSOVITCH, N. (1953). *Anthropologie et groupes sanguins des populations du Maroc*. Masson et Cie, Paris.

KOYOUMDJISKY-KAYE, E., ZILBERMAN, Y., & ZEEVI, Z. (1976). A comparative study of tooth and dental arch dimensions in Jewish children of different ethnic descent. i. Kurds and Yemenites. *Amer. J. phys. Anthrop.* **44**, 437–43.

KRAIN, L. S., TERASAKI, P. I., NEWCOMER, V. D., & MICKEY, M. R. (1973). Increased frequency of HL-A10 in pemphigus vulgaris. *Arch. Derm.* **108**, 803–5.

KRIKLER, D. M. (1969). Diabetes in Rhodesian Sephardic Jews. *S. Afr. med. J.* **43**, 931–3.

KRIKLER, D. M. (1970). Diseases of Jews. *Postgrad. med. J.* **46**, 687–97.

LAHAV, M. & SZEINBERG, A. (1972). Red-cell glutamic-pyruvic transaminase polymorphism in several population groups in Israel. *Hum. Hered.* **22**, 533–8.

LANDSTEINER, K. & LEVINE, P. (1927). Further observations on individual difference of human blood. *Proc. Soc. exp. Biol., N.Y.* **24**, 941–2.

LANDSTEINER, K. & WIENER, A. S. (1940). An agglutinable factor in human blood recognized by immune sera for rhesus blood. *Proc. Soc. exp. Biol., N.Y.* **43**, 223.

LASCH, E. E., RAMOT, B., & NEUMANN, G. (1968). Childhood celiac disease in Israel. *Israel J. med. Sci.* **4**, 1260–4.

LASKER, M., ENKLEWITZ, M., & LASKER, G. W. (1936). The inheritance of l-xyloketosuria (essential pentosuria). *Hum. Biol.* **8**, 243–55.

LAVRIK, S. S., KARAVANOV, A. G., DANILOVA, E. I., DUDNIK, M. I., & DYACHENKO, V. D. (1968). The distribution of blood antigens of the system ABO, MN, Rh among the population of the Ukraine in connection with some problems of ethnic history. *8th Int. Congr. anthrop. ethnol. Sci.*, Tokyo and Kyoto, **1**, 200–3.

LEAKEY, T. E. B., COWARD, A. R., WARLOW, A., & MOURANT, A. E. (1972). The distribution in human populations of electrophoretic variants of cytoplasmic malate dehydrogenase. *Hum. Hered.* **22**, 542–51.

LEBLANC, M. (1946). Note sur la sérologie des populations juives du Maroc. *Bull. Inst. Hyg. Maroc.* n.s. **6**, 5–14.

LEHMANN, H. & HUNTSMAN, R. G. (1974). *Man's haemoglobins. Including the haemoglobinopathies and their investigation.* North-Holland, Amsterdam and Oxford.

LEICHIK, M. S. (1928). Die Blutgruppen und die Untersuchung der Vererbung der agglutinierenden Substanzen bei den übergesiedelten Hebräern des Odessaer Gebiets. *Ukr. Zbl. Blutgr.* **2**, 36–44.

LEVENE, C. (1968). Pers. comm. to A. E. Mourant.

LEVENE, C. (1975). Pers. comm. to A. E. Mourant.

LEVENE, C. & COHEN, T. (1977). Gc in Kurdish Jews. Pers. comm. to A. E. Mourant.

LEVENE, C., MEDALIE, J. H., & COHEN, T. (1977a). Haptoglobin and Gc in Polish and Iraq Jews. Pers. comm. to A. E. Mourant.

LEVENE, C., SIMHAI, B., SELA, R., & COHEN, T. (1977b). Blood groups of Iranian (Persian) Jews; (A_1A_2BO, CDEce, MNSs, Kk, Fy(a)). Pers. comm. to A. E. Mourant.

LÉVÊQUE, J. (1955). Les groupes sanguins des populations marocaines. *Bull. Inst. Hyg. Maroc.* **15**, 237–321.

LEVINE, P., BACKER, M., WIGOD, M., & PONDER, R. (1949). A new human hereditary blood property (Cellano) present in 99·8% of all bloods. *Science* **109**, 464–6.

LEVINE, P., BOBBITT, O. B., WALLER, R. K., & KUHMICHEL, A. (1951). Isoimmunization by a new blood factor in tumor cells. *Proc. Soc. exp. Biol.* **77**, 403–5.

LIVINGSTONE, F. B. (1967). *Abnormal hemoglobins in human populations. A summary and interpretation.* Aldine Publishing Company, Chicago.

LÉVY, J., NICOLI, R. M., & RANQUE, J. (1967). Études seroanthropologiques. xxiii. Le peuplement israélite dans le sud-est de la France. *Biométr. hum.* **1**, 172–80.

LEWIS, M., CHOWN, B., & PETERSON, R. F. (1955). On the Kell-Cellano (K-k) blood group. The distribution of its genes in the white population of Manitoba. *Amer. J. phys. Anthrop.* **13**, 323–30.

LEWIS, M., KAITA, H., & CHOWN, B. (1972). The Duffy blood group system in Caucasians. *Vox Sang.* **23**, 523–7.

LEWIS, W. H. P. & HARRIS, H. (1967). Human red cell peptidases. *Nature, Lond.* **215**, 221–3.

LUBINSKI, H., BENJAMIN, B., & STREAN, G. J. (1945). The distribution of the Rh factor in Jewish mothers and infants and the incidence of haemolytic anaemia in Jewish newborn infants. *Can. med. Ass. J.* **53**, 28–30.

LUNDSGAARD, A. & JENSEN, K. G. (1971). Pers. comm. to A. E. Mourant.

MACFARLANE, E. W. E. (1937). The racial affinities of the Jews of Cochin. *J. roy. Asiat. Soc. Bengal.* **3**, 1–23.

McKUSICK, V. A. (1966). *Mendelian inheritance in man. Catalogs of autosomal dominant, autosomal recessive,*

and X-linked phenotypes. William Heinemann Medical Books Ltd., London.

McKUSICK, V. A., NORUM, R. A., FARKAS, H. J., BRUNT, P. W., & MAHLOUDJI, M. (1967). The Riley–Day syndrome—observations on genetics and survivorship. *Israel J. med. Sci.* **3**. (Reprinted in Shiloh & Selavan 1973).

MACMAHON, B. & FOLUSIAK, J. C. (1958). Leukemia and ABO blood groups. *Amer J. hum. Genet.* **10**, 287–93.

MANN, J. D., CAHAN, A., GELB, A. G., FISHER, N., HAMPER, J., TIPPETT, P., SANGER, R., & RACE, R. R. (1962). A sex-linked blood group. *Lancet* **i**, 8–10.

MANUILA, A., SAUTER, M. R., & VESTEMEANU, M. (1945). Étude de 16.685 corrélations entre le groupe sanguin et d'autres caractères morphologiques, examinés en Europe orientale. *Annexe Arch. suisses Anthrop. gén.* **1**, 47–107.

MANUILA, S. (1924). Recherches séro-anthropologiques sur les races en Roumanie par la méthode de l'isohémagglutination. *C.R. Soc. Biol., Paris* **90**, 1071–3.

MARGOLIS, E., GUREVITCH, J., & HASSON, E. (1957). Blood groups in Jews from Morocco and Tunisia. *Ann. hum. Genet.* **22**, 65–8.

MARGOLIS, E., GUREVITCH, J., & HERMONI, D. (1960). Blood groups in Ashkenazi Jews. *Amer. J. phys. Anthrop.* **18**, 201–3.

MARGOLIS, E., GUREVITCH, J., & HERMONI, D. (1960). Blood groups in Sephardic Jews. *Amer. J. phys. Anthrop.* **18**, 197–9.

MARGOLIS, E. (1962). A new hereditary syndrome—sex-linked deaf-mutism associated with total albinism. *Acta genet. Stat. med.* **12**, 12–19.

MECHALI, D., LÉVÊQUE, J., & FAURE, P. (1957). Les groupes sanguins ABO et Rh des Juifs du Maroc. *Bull. Soc. Anthrop., Paris* **8**, 354–70.

MELKIKH & GRINGOT (1926). Quoted by JETTMAR, H. M. (1930).

MICLE, S., KOBILYANSKY, E. D., NATHAN, M., & MOR, S. (1978). Serum alkaline phosphatase phenotypes and secretor status in several Jewish populations in Israel. *Hum. Hered.* in press.

MILLER, L. H., MASON, S. J., DVORAK, J. A., McGINNISS, M. H., & ROTHMAN, I. K. (1975). Erythrocyte receptors for (*Plasmodium knowlesi*) malaria: Duffy blood group determinants. *Science* **189**, 561–3.

MODRZEWSKA, K. (1958). Wrażliwość ludności polskiej na fenyltiokarbamid. (Sensitivity to phenylthiocarbamide in the Polish population.) *Przegl. antrop.* **24**, 540–64. (In Polish).

MONCEAUX, P. (1970). Les colonies juives dans l'Afrique romaine. *Cah. Tunis.* **18**, 157–84.

MOULLEC, J., & ABDELMOULA, H. (1954). Quelques données sur les groupes sanguins des Tunisiens. *Sem. Hôp. Paris.* **30**, 3061–2.

MOURANT, A. E. (1946). A 'new' human blood group antigen of frequent occurrence. *Nature, Lond.* **158**, 237.

MOURANT, A. E., KOPEĆ, A. C., & DOMANIEWSKA-SOBCZAK, K. (1958). *The ABO blood groups. Comprehensive tables and maps of world distribution.* Blackwell Scientific Publications, Oxford; Charles C. Thomas, Springfield, Ill.; The Ryerson Press, Toronto. Reprinted as an Appendix to Mourant *et al.* 1976.

MOURANT, A. E. (1959). The blood groups of the Jews. *Jewish J. Sociol.* **1**, 155–76.

MOURANT, A. E., KOPEĆ, A. C., & DOMANIEWSKA-SOBCZAK, K. (1976a). *The distribution of the human blood groups and other polymorphisms.* (2nd edn.) Oxford University

Press. *Appendix*: Reprint of Mourant *et al.* 1958, *The ABO blood groups*.

MOURANT, A. E., TILLS, D., & DOMANIEWSKA-SOBCZAK, K. (1976b). Sunshine and the geographical distribution of the alleles of the Gc system of plasma proteins. *Humangenetik*, **33**, 307–14.

MOURANT, A. E., KOPEĆ, A. C., & DOMANIEWSKA-SOBCZAK, K. (1978). *Blood groups and diseases*. Oxford University Press. (Oxford monographs on medical genetics.)

MUHSAM, H. V. (1964). The genetic origin of the Jews. *Genus* **20**, 3–30 in reprint.

MYRIANTHOPOULOS, N. C. & ARONSON, S. M. (1966). Population dynamics of Tay-Sachs disease. I. Reproductive fitness and selection. *Amer. J. hum. Genet.* **18**, 313–27.

NEEL, J. V. (1962). Diabetes mellitus: a 'thrifty' genotype rendered detrimental by 'progress'? *Amer. J. phys. Anthrop.* **14**, 353–60.

NEUMAN, J., NOVIZKI, I., BAUERBERG, J., & STEINBERG, J. (1961). Distribución de frecuencias de los grupos sanguíneos del sistema ABO en los enfermos con infarto de miocardio. *Rev. Asoc. méd. argent.* **75**, 534–40.

NISENBAUM, C., SANDBANK, U., & KOHN, R. (1964). Pezizaeus-Merzbacher disease 'infantile acute type'. Report of a family. *Ann. Paediat.* **204**, 365–76.

NYHAN, W. L., BORDEN, M., & CHILDS, B. (1961). Idiopathic hyperglycinemia: a new disorder of amino acid metabolism. II. The concentrations of the amino acids in the plasma and their modification by the administration of leucine. *Pediatrics* **27**, 539–50.

OTTENSOOSER, F., LEON, N., SATO, M., & SALDANHA, P. H. (1962). O fator Diego e outros grupos sanguíneos em judeus ashkenazim. *14th Reun. An. Soc. Brasil. Progr. Ciên.*, Curitiba. Resumos de Comunicações, p. 25.

OTTENSOOSER, F., LEON, N., SATO, M., & SALDANHA, P. H. (1963). Blood groups of a population of Ashkenazi Jews in Brazil. *Amer. J. phys. Anthrop.* **21**, 41–8.

OXFORD BIBLE ATLAS (1962). Edited by H. G. May, R. N. Hamilton, and G. N. S. Hunt. Oxford University Press.

PALATNIK, M. (1965). Distribution of the Di^a factor in Argentine Jews. *Nature, Lond.* **207**, 1203–4.

PALMARINO, R., AGOSTINO, R., GLORIA, F., LUCARELLI, P., BUSINCO, L., ANTOGNONI, G., MAGGIONI, G., WORKMAN, P. L., & BOTTINI, E. (1975). Red cell acid phosphatase: another polymorphism correlated with malaria? *Amer. J. phys. Anthrop.* **43**, 177–85.

PARR, L. W. (1931). Blood studies of peoples of Western Asia and North Africa. *Amer. J. phys. Anthrop.* **16**, 15–29.

PARR, L. W. (1934). Taste blindness and race. *J. Hered.* **25**, 187–90.

PATAI, R. (1953). *Israel between East and West. A Study in human relations*. The Jewish Publication Society of America, Philadelphia.

PATAI, R. & PATAI WING, J. (1975). *The myth of the Jewish race*. Charles Scribner's Sons, New York.

PAULING, L., ITANO, H. A., SINGER, S., & WELLS, I. C. (1949). Sickle cell anaemia, a molecular disease. *Science* **110**, 43–8.

PETROV, G. I. (1928). Zur Frage der Isoagglutination des Blutes bei den Tadjiken. *Z. Rassenphysiol.* **1**, 85–9.

PEVZNER. Quoted by RUBASHKIN, V. Ya. (1929).

POLO, M. (1959). *The travels of Marco Polo*. Translated and with an introduction by R. Latham. Penguin Books, Harmondsworth.

POST, R. H. (1965). Jews, genetics and disease. *Eugen. Quart.* **12**, 162–6.

POTAPOV, M. I. (1970). Obnaruzheniye antigena sistemy Lewis, svoystvennogo eritrotsitam vydeliteley gruppy Le(a-b-). (Detection of the antigen of the Lewis system, characteristic of the erythrocytes of the secretory group Le(a-b-).) *Probl. Gemat. Pereliv. Krovi* **15**, no. 11, 45–9.

PRINZ, J. (1974). *The secret Jews*. Vallentine, Mitchell, London.

PULYANOS, A. N. (1963). K antropologii karaimov Litvy i Kryma. (On the anthropological characteristics of the Karaimes of Lithuania and Crimea.) *Vopr. Antrop.* **13**, 116–33.

RACE, R. R. (1944). An 'incomplete' antibody in human serum. *Nature, Lond.* **153**, 771–2.

RACE, R. R. & SANGER, R. (1975). *Blood groups in man*. 6th ed. Blackwell Scientific Publications, Oxford.

RACZ, I., MOLCHO, M., SHAPIRO, B., BUBIS, J., & BAUM, G. L. (1964). Cystic disease of lung. *Proc. Tel Hashomêr Hosp.* **3**, 187–92.

RAKOVSKY & SUKHOTIM. Quoted by SEMENSKAYA, E. M. (1930).

RAMOT, B., ZIKERT-DUVDEVANI, P., & TAUMAN, G. (1961). Distribution of haptoglobin types in Israel. *Nature, Lond.* **192**, 765–88.

RAMOT, B., DUVDEVANI-ZIKERT, P., & KENDE, G. (1962). Haptoglobin and transferrin types in Israel. *Ann. hum. Genet.* **25**, 267–71.

RAMOT, B., ADAM, A., BONNÉ, B., GOODMAN, R. M., & SZEINBERG, A. (eds.) (1974). *Genetic polymorphisms and diseases in man*. Sheba Int. Symposium, Tel Aviv, 1973. Academic Press, New York, London.

RANQUE, J., NICOLI, R. M., GHERIB, B., & BATTAGLINI, P. F. (1964). Étude séro-anthropologique des Juifs de l'île de Djerba (Tunisie). *9th Congr. int. Soc. Blood Transfus.*, Mexico, 1962. *Bibl. haemat.* **19**, 210–12.

RAPLEY, S., ROBSON, E. B., HARRIS, H., & MAYNARD SMITH, S. (1967). Data on the incidence, segregation and linkage relations of the adenylate kinase (AK) polymorphism. *Ann. hum. Genet.* **31**, 237–42.

RASKINA, R. I. (1930). Ueber die Verteilung der Blutgruppen bei einigen Arten von Geisteskranken. *Ukr. Zbl. Blutgr.* **4**, 192–4.

RAUSEN, A. R., ROSENFIELD, R. E., ALTER, A. A., HAKIM, S., GRAVEN, S. N., APOLLON, C. J., DALLMAN, P. R., DALZIEL, J. C., KONUGRES, A. A., FRANCIS, B., GAVIN, J., & CLEGHORN, T. E. (1967). A 'new' infrequent red cell antigen, Rd (Radin). *Transfusion, Philad.* **7**, 336–42.

REICHER, M. (1932). Sur les groupes sanguins des Caraïmes de Troki et de Wilno. *Anthropologie, Prague* **10**, 259–67.

RIFE, D. C. & SCHONFELD, M. D. (1944). A comparison of the frequencies of certain genetic traits among Gentile and Jewish students. *Hum. Biol.* **16**, 172–80.

RIFE, D. C. (1957). Pers. comm. to A. E. Mourant.

ROBERTS, J. A. Fraser (1957–8). Pers. comm.

ROBSON, E. B. & HARRIS, H. (1966). Further data on the incidence and genetics of the serum cholinesterase phenotype C_5+. *Ann. hum. Genet.* **29**, 403–8.

ROSENBLOOM, J. R. (1966a). Moroccan Jewry: a community in decline. *Judaism* **15**, 217–22.

ROSENBLOOM, J. R. (1966b). A note on the size of the Jewish communities in the south of Morocco. *Jewish J. Sociol.* **8**, 209–12.

ROSENTHAL, R. L., DRESKIN, O. H., & ROSENTHAL, N. (1955). Plasma thromboplastin antecedent (PTA) deficiency. *Blood* **10**, 120–31.

ROSENTHAL, R. L. (1964). Haemorrhage in PTA (Factor XI)

deficiency. *10th Congr. Int. Soc. Haemat.*, Stockholm (Abstract).

ROTEM, Y. (1970). Gene-frequencies in Jews. *Lancet* ii, 315.

ROTH, C. (1969). *A short history of the Jewish people*. East and West Library, London.

ROUTIL, R. (1933). Über die Wertigkeit der Blutgruppen-befunde in Vaterschaftsprozessen. *Z. Rassenphysiol.* **6**, 70–4.

ROZEN, P. & SHAFRIR, E. (1968). Behaviour of serum free fatty acids and glucose during lactose tolerance tests. *Israel J. med. Sci.* **4**, 100–9.

RUBASHKIN, V. Ya. & DERMAN. Quoted by HIRSZFELD, L. (1928).

RUBASHKIN, V. Ya. (1929). *Krovyanye gruppy*. (Blood groups) Gosudarstvennoye izdatel'stvo, Moskva, Leningrad.

RYDER, L. P. & SVEJGAARD, A. (1976). *Associations between HLA and disease*. Report from the HLA and Disease Registry of Copenhagen. [Reprinted as Appendix to Mourant *et al. Blood groups and diseases.*]

SABOLOTNY, S. S. (1928). Die Blutgruppen der Karaimen und Krimtschaken. *Ukr. Zbl. Blutgr.* **3**, 12–22.

SACHS, L. & BAT-MIRIAM, M. (1957). The genetics of Jewish populations. I. Finger print patterns in Jewish populations in Israel. *Amer. J. hum. Genet.* **9**, 117–26.

SAKHAROV (1930). Quoted by STEFFAN, P. & WELLISCH, S. (1933).

SALDANHA, P. H. & BEÇAK, W. (1959). Taste thresholds for phenylthiourea among Ashkenazic Jews. *Science* **129**, 150–1.

SANGER, R. (1955). An association between the P and Jay systems of blood groups. *Nature, Lond.* **176**, 1163–4.

SANGER, R., RACE, R. R., & JACK, J. (1955). The Duffy blood groups of the New York Negroes: the phenotype Fy (a-b-). *Brit. J. Haemat.* **1**, 370–4.

SAUGSTAD, L. F. (1973). Association between phenylketonuria gene and Rhesus, P.G.M. and Kell systems. *Lancet* ii, 1258.

SAUGSTAD, L. F. (1975). Frequency of phenylketonuria in Norway. *Clin. Genet.* **7**, 40–51.

SAY, B., OZAND, P., BERKEL, I., & ÇEVIK, N. (1965). Erythrocytose glucose-6-phosphate dehydrogenase deficiency in Turkey. *Acta paediat. scand. (Uppsala).* **54**, 319–24.

SCHANFIELD, M. S., GILES, E., & GERSHOWITZ, H. (1975). Genetic studies in the Markham Valley, northeastern Papua New Guinea: gamma globulin (Gm and Inv), group specific component (Gc) and ceruloplasmin (Cp) typing. *Amer. J. phys. Anthrop.* **42**, 1–7.

SCHIFF, F. & ZIEGLER. Quoted by STEFFAN, P. & WELLISCH, S. (1932).

SCHIFF, F. (1940). Racial differences in frequency of the 'secreting factor'. *Amer. J. phys. Anthrop.* **27**, 255–62.

SELIGSOHN, U., SHANI, M., RAMOT, B., ADAM, A., & SHEBA, C. (1970). Dubin-Johnson syndrome in Israel. II. Association with Factor VII deficiency. *Quart. J. Med.* **39**, 569–84.

SELTZER, C. C. (1940). *Contributions to the racial anthropology of the Near East*. Peabody Museum, Cambridge, Mass.

SEMENSKAYA, E. M. (1930). Krovyanye gruppy u gruzinskikh yevreyev. (Blood groups in Georgian Jews). *Zh. Pat. Yevreyev* **1**, 38.

SEMENSKAYA, E. M., et al. (1937a). Pers. comm. to A. E. Mourant.

SEMENSKAYA, E. M., et al. (1937b). Quoted by BOYD, W. C. & BOYD, L. G. (1937). *Amer. J. phys. Anthrop.* **23**, 49–70,

and supplemented by pers. comm. from W. C. Boyd to A. E. Mourant.

SHANI, M., SELIGSOHN, U., GILON, E., SHEBA, C., & ADAM, A. (1970). Dubin-Johnson syndrome in Israel. I. Clinical, laboratory and genetic aspects of 101 cases. *Quart. J. Med.* **39**, 549–68.

SHAPIRO, H. L. (1960). *The Jewish people: a biological history*. UNESCO.

SHATSKAYA, T. L., KRASNOPOLSKAYA, K. D., & IDELSON, L. J. (1976). Mutant forms of erythrocyte glucose-6-phosphate dehydrogenase in Ashkenazi. Description of two new variants: G6PD Kirovograd and G6PD Zhitomir. *Hum. Genet.* **33**, 175–8.

SHEBA, C., SZEINBERG, A., & RAMOT, B. (1961). Distribution of glucose-6-phosphate dehydrogenase deficiency (G6PD) among various communities in Israel. *2nd Int. Congr. hum. Genet.*, Rome, 633–4.

SHEBA, C., ASHKENAZI, I., & SZEINBERG, A. (1962). Taste sensitivity to phenylthiourea among the Jewish population groups in Israel. *Amer. J. hum. Genet.* **14**, 44–51.

SHEBA, C., SZEINBERG, A., RAMOT, B., ADAM, A., & ASHKENAZI, I. (1962). Epidemiologic surveys of deleterious genes in different population groups in Israel. *Amer. J. publ. Hlth.* **52**, 1101–6.

SHEBA, C. (1970). Gene frequencies in Jews. *Lancet* i, 1230–1.

SHEBA, C. (1971). Jewish migration in its historical perspective. *Israel J. med. Sci.*, **7**, 1333–41.

SHILOH, A. & COHEN SELAVAN, I. (eds.) (1973). *Ethnic groups of America: their morbidity, mortality and behavior disorders.* Vol. I *The Jews*. Charles S. Thomas, Springfield, Ill. [Collection of reprints.]

SHIRYAK, E. A. (1929). Blutgruppen in Cherson. *Ukr. Zbl. Blutgr.* **3**, (Abstract in *Z. Rassenphysiol.* 1931 **4**, 46).

SHOKEID, M. (1971). *The dual heritage: Immigrants from the Atlas mountains in an Israeli village.* Manchester University Press.

SILBERSTEIN, W. & GOLDSTEIN, N. (1958). Blood groups in women of various oriental Jewish communities. (In Hebrew). *Harefuah* **54**, 295–6.

SIRSAT, S. E. (1956). Effect of migration on some genetical characters in six endogamous groups in India. *Ann. hum. Genet.* **21**, 145–54.

ŠKALOUD (1934). Quoted by STEFFAN, P. & WELLISCH, S. (1936).

SLOUSCHZ, N. (1927). *Travels in North Africa*. Jewish Publ. Soc. of America, Philadelphia.

SMIRNOVA & CHERNYAEVA (1929). Quoted by STEFFAN, P. & WELLISCH, S. (1932).

SMITHIES, O., CONNELL, G. E., & DIXON, G. H. (1962). Inheritance of haptoglobin subtypes. *Amer. J. hum. Genet.* **14**, 14–21.

SOLAL, R. & HANOUN, W. (1952). Pers. comm. to A. E. Mourant.

SPENCER, N., HOPKINSON, D. A., & HARRIS, H. (1964). Phosphoglucomutase polymorphism in man. *Nature, Lond.* **204**, 742–5.

SPENCER, N., HOPKINSON, D. A., & HARRIS, H. (1968). Adenosine deaminase polymorphism in man. *Ann. hum. Genet.* **32**, 9–14.

STARK, G. J. & HENNER, J. (1953). Hematologic values in blood donors and the ABO blood distribution in Tel Aviv. (In Hebrew). *Harefuah* **45**, 175–6.

STEFFAN, P. & WELLISCH, S. (1932). Die geographische Verteilung der Bluteigenschaften O, A und B. In *Hand-*

buch der Blutgruppenkunde (ed. P. Steffan), pp. 396–433. Lehmanns, Munich.

STEFFAN, P. & WELLISCH, S. (1933 & 1936). Die geographische Verteilung der Blutgruppen. *Z. Rassenphysiol.* **6**, 28–36 & **8**, 38–47.

STEINBERG, A. G., LEVENE, C., GOLDSCHMIDT, E., & COHEN, T. (1970). The Gm and Inv allotypes in kindreds of Kurdish Jews. *Amer. J. hum. Genet.* **22**, 652–61.

STEINBERG, A. G. (1973). The Gm and Inv allotypes of some Ashkenazic Jews living in Northern U.S.A. *Amer. J. phys. Anthrop.* **39**, 409–11.

STERN, C. (1973). *Principles of human genetics.* (3rd edn.), W. H. Freeman & Co, San Francisco.

STONE, M. E. (1973). Judaism at the time of Christ. *Sci. Amer.* **228**, 80–7.

STRIZOWER, S. (1971). *The Bene Israel of Bombay.* Schocken Books, New York.

STROUP, M., MacILROY, M., WALKER, R., & AYDELOTTE, J. V. (1965). Evidence that Sutter belongs to the Kell blood group system. *Transfusion, Philad.* **5**, 309–14.

SVEJGAARD, A., HAUGE, M., JERSILD, C., PLATZ, P., RYDER, L. P., NIELSEN, L. S., & THOMSEN, M. (1975). *The HLA system. An introductory survey.* S. Karger, Basel.

SZEINBERG, A. (1963). G6PD deficiency among Jews—genetic and anthropological considerations. In *The genetics of migrant and isolate populations* (ed. Elizabeth Goldschmidt), pp. 69–72. Williams and Wilkins, Baltimore.

SZEINBERG, A., BAR-OR, R., & SHEBA, C. (1961). Distribution of isoniazid inactivator types in various Jewish groups in Israel. *2nd Int. Congr. hum. Genet.*, Rome, 110–12.

SZEINBERG, A., PIPANO, S., & OSTFELD, E. (1966). Frequency of atypical pseudocholinesterase in different population groups in Israel. *2nd Congr. europ. Anaesth.*, Kopenhagen. *Acta anaesth.*, 1966, suppl. 24, 199–205.

SZEINBERG, A., PIPANO, S., ROZANSKY, Z., & RABIA, N. (1971). Frequency of red cell adenosine deaminase phenotypes in several population groups in Israel. *Hum. Hered.* **21**, 357–61.

SZEINBERG, A. & TOMASHEVSKY-TAMIR, S. (1971). Red cell adenylate kinase and phosphoglucomutase polymorphisms in several population groups in Israel. *Hum. Hered.* **21**, 289–96.

SZEINBERG, A., PIPANO, S., ROZANSKY, Z., & RABIA, N. (1972). High frequency of atypical pseudocholinesterase gene among Iraqi and Iranian Jews. *Clin. Genet.* **3**, 123–7.

TALMON, S. (1977). The Samaritans. *Sci. American* **236**, 100–8.

TERRY, M. C., & SEGALL, G. (1947). The association of diabetes and taste-blindness. *J. Hered.* **38**, 135–7.

THOMSEN, O., FRIEDENREICH, V., & WORSAAE, E. (1930). Uber die Möglichkeit der Existenz zweier neuer Blutgruppen; auch ein Beitrag zur Beleuchtung sogenannter Untergruppen. *Acta path. microbiol. scand.* **7**, 157–90.

TILLS, D., WARLOW, A., MOURANT, A. E., KOPEĆ, A. C., EDHOLM, O. G., & GARRARD, G. (1977). The blood groups and other hereditary blood factors of the Yemenite and Kurdish Jews. *Ann. hum. Biol.* **4**, 259–74.

TIPPETT, P. (1967). Genetics of the Dombrock blood group system. *J. med. Genet.* **4**, 7–11.

TROKAN (1929). Quoted by BOYD, W. C. (1939).

UNDEVIA, J. V. (1969). *Population genetics of the Parsis. Comparison of genetical characteristics of the present Parsi population with its ancestral and affiliated groups.* Ph.D. Thesis, Bombay.

VISHNEVSKY. Quoted by PETROV, G. T. (1928).

VRIES, A. DE, JOSHUA, H., LEHMANN, H., HILL, R. L., & FELLOWS, R. E. (1963). The first observation of an abnormal haemoglobin in a Jewish family. Haemoglobin Beilinson. *Brit. J. Haemat.* **9**, 484–6.

WAGNER, L. B. (1926). Svodka raspredeleniya krovyanykh grupp sredi naseleniya zemnogo shara k 1926 g. (Global distribution of blood groups by 1926). *Russk. antrop. Zh.* **15**, no. 1–2, 75–88.

WALKER, R. H., ARGALL, C. I., STEANE, E. A., SASAKI, T. T., & GREENWALT, T. J. (1963). Js^b of the Sutter blood group system. *Transfusion, Philad.* **3**, 94–9.

WEINER, J. S. & HUIZINGA, J., (eds.) (1972). *The assessment of population affinities in man.* Clarendon Press, Oxford.

WELSH, J. D., ZSCHIESCHE, O. M., WILLITS, V. L., & RUSSELL, L. (1968). Studies of lactose intolerance in families. *Arch. intern. Med.* **122**, 315–17.

WHITE, W. C. (1966). *Chinese Jews. A compilation of matters relating to Jews of K'ai-Feng Fu.* (2nd edn.). University of Toronto Press.

WIENER, A. S. *et al.* (1929). Quoted by STEFFAN, P. & WELLISCH, S. (1932).

WOLMAN, M., STERK, V. V., GATT, S., & FRENKEL, M. (1961). Primary familial xanthomatoses with involvement and calcification of the adrenals. *Pediatrics* **28**, 742–57.

WORLD HEALTH ORGANIZATION (1967). *Manual of the international statistical classification of diseases, injuries, and causes of death.* WHO, Geneva.

ỸOUNOVITCH, R. (1932). Contribution à l'ètude sérologique des Juifs de Yémen. *C.R. Soc. Biol., Paris* **111**, 929–31.

ỸOUNOVITCH, R. (1933a). Étude sérologique des Juifs samaritains. *C.R. Soc. Biol., Paris* **112**, 970–1.

ỸOUNOVITCH, R. (1933b). Les caractères sérologiques des Juifs asiatiques. *C.R. Soc. Biol., Paris* **113**, 1101–3.

ỸOUNOVITCH, R. (1938). Quoted by GENNA, G. E. (1938).

ZIPRKOWSKI, L., KRAKOWSKI, A., ADAM, A., COSTEFF, H., & SADE, J. (1962). Partial albinism and deaf mutism. *Arch. Derm.* **86**, 530–9.

ZIPRKOWSKI, L. & ADAM, A. (1964). Recessive total albinism and congenital deaf-mutism. *Arch. Derm.* **89**, 151–5.

ZIPRKOWSKI, L. & SCHEWACH-MILLET, M. (1964). A long-term study of pemphigus. *Proc. Tel-Hashomer Hosp.* **3**, 46–53.

ZU RHEIN, G. M., EICHMAN, P. L., & PULETTI, F. (1960). Familial idiocy with spongy degeneration of the central nervous system of Van Bogaert-Bertrand type. *Neurology* **10**, 998–1006.

INDEX TO THE TEXT

This index covers only the text of this book; the tables are indexed separately (p. 119). A large proportion of the entries concern Jewish populations and their genetics. These are indexed primarily under their ethnic (e.g. 'Sephardim') or topographic (e.g. 'Yemenite') descriptions, not under the word 'Jews'. Under each of these population headings separate entries are included for such features as blood groups, other polymorphisms, and prevalent congenital diseases.

INDEX TO TABLES

119